RETURN TO HARMONY

Creating harmony and balance
through the frequencies of sound.

Nicole LaVoie

This book is not intended to treat, diagnose or prescribe. The information contained herein is in no way to be considered as a substitute for your own inner guidance, or consultation with a duly licensed health care professional.

Cover design: Francois Lapierre
Re-writing and refining text: Jill Lawrence
Editing: Debbie Darling
Page composition and book design: Susan Tinkle
Illustrations: Helene Gauthier

First published in 1996 by:
  Sound Wave Energy Press
  P.O. Box 3969
  Pagosa Springs, CO 81147

Second Edition, 1998
Printed in the United States of America

Library of Congress Cataloging in Publication Data
La Voie, Nicole,
    Return to Harmony: Creating harmony and balance
    through the frequencies of sound.
    Catalog card # 96-92084
    ISBN 0-9650387-1-8

RETURN TO HARMONY

Creating harmony and balance
through the frequencies of sound.

Second Edition

NICOLE LAVOIE

Table of Contents

Acknowledgments

I offer my gratitude and appreciation to all the lovely people who have shared their love and opened their hearts to me during the journeys of my life, and especially to those who helped make this book a reality.

My sincere appreciation and special thanks to Jill Lawrence for her outstanding accomplishment in re-writing this book with such elegance, while respecting its essence and its style.

To my son Robert who supports the work by producing all of the Sound Wave Energy tapes. His impeccable standards result in the tapes having the true frequencies that are required, as each one has to be produced in "real-time" and cannot be duplicated.

To my son Francois who produced my logo and customized many special programs, who supported my early beginnings to befriend my computer, and continues to provide wonderful technical support.

Special thanks to Debbie Darling with her precious editing skills, to Annabelle and Bob Metteer with their participation in putting the manuscript together; to Helene Gauthier, the artist who created the illustrations which beautifully enlighten "Return to Harmony," and to Susan Tinkle for the beautiful book design, page composition, and final editing of the second edition.

Thank you to all the distributors of the SWE tapes, and all that want to learn to do SoulNote/SWE—they are the ones that will spread the news of SWE far and wide. To all the people that have used the tapes and shared their results with me and/or others. Their enthusiasm has truly supported my work.

To my guides and angels that have supported me all the way, as I return to harmony.

Foreword

by Roy B. Kupsinel, M.D.

You may be asking, "Who is Nicole LaVoie?" She is a dynamic lady, a ball of radiating, energy-beaming light, unconditional love, joy, happiness, health, peace and freedom to all those who meet her. She's a petite, middle-aged French Canadian with a charming French accent enhanced by an almost perpetual smile.

Nicole has a college background and twenty years of hospital experience as an x-ray technician. She has taken courses to prepare her for the field of Sound Wave Energy (SWE). Nicole is a Reiki Master, has a background in psychology and philosophy of religion, and does hands-on healing, including work with crystals, magnets, and frequency techniques. She has been involved with energy research since 1974 and founded SWE in 1993 as a result of studies of musical sound and its resonance. Nicole is a true servant of humanity and is divinely guided not only with her energy work but also with her travels about the planet.

Next question, "What are SWE tapes?" Let me tell you what SWE is all about! Sound waves have resonance. Resonance is the frequency at which a person or an object vibrates. All objects in the universe have a frequency. To quote from Nicole's SWE booklet: "In the beginning was the Word. Word is sound and sound is The Way."

The purpose of these tapes is to bring the body into a state of balance and harmony. In my thirty-three years of medical practice, SWE tapes are one of the most powerful holistic tools I have encountered. That's a very powerful statement that expresses exactly how I feel!

Introduction

The following goals have drawn this book out of my head and heart and onto these pages.

First, I desire effective duplication of my spoken words. I say the same things repeatedly to the wonderful people I meet. It is my hope that these written words will become "clones" of my spoken words, that they will reach more people and that they will free the time for me to do more research and to play more.

Secondly, I want a clear, complete, and well-paced description of Sound Wave Energy (SWE). When I give oral descriptions or lectures, there is not always time to be as complete as I'd like to be. Also, I sometimes simply forget to share an important thing that has become very familiar to me.

My third goal is a desire to establish a warm and personal connection with those "listening" to me on these pages. While you "hold me" in your hands through this book, it may well be more personal than a lecture.

The SWE products and services that I provide for people have their origins in my life story. My biographical stories have moved many people to tears of recognition and joy, to laughter, and to "aha's" that they value. So, a fourth goal is to ensure that many more people can learn of such benefits than would be the case if I merely continued to tell my stories in person.

Also, embedded in the many stories and testimonials is technical education about the effective modality of SWE. Therefore, a powerful fifth goal for me is to avoid writing dry and sterile text. I prefer instead the real stories of real discoveries and real feelings of real people.

As well as writing about the history and current status of SWE, I want to invite you to co-create its future. So a sixth goal for this book is an invitation for you to give the input, support and participation you may feel like adding to the dramatic unfoldment that's occurring for Sound Wave Energy.

The underlying goal of *Return to Harmony* is, of course, "the greatest good for the greatest number." What structures and functions will we co-create to get the benefits of SWE to the greatest number of people?

I have been told that it may help you to "hear" that these words written in English have a strong French-Canadian accent. English is my second language. Since the book is coming from my heart, I ask you to read it with your heart. Now, may we enjoy *Return to Harmony* together.

Nicole La Voie
December, 1995
Pagosa Springs, Colorado

Part One

The Narrative

Early Lessons in Life

Girlhood

We all learn our life lessons in different ways. We are each presented with the ideal circumstances for our ideal learning. I was no exception. For example, I learned about injustice in a fairly unique way. Instead of being downtrodden and discriminated against, I was the favored one, the one who could do no wrong.

I was born in Canada, the eighth child in a French Canadian family of nine children. I was blessed with parents who gave me large amounts of attention and support, which I translated as love. When I was very young, I learned what it felt like to have all my desires fulfilled—so much so that I began to feel the benefits I enjoyed were an injustice to my brothers and sisters. I noticed that they had to make strong demands to have some of their desires fulfilled by our parents; however, I received whatever I wanted through mere suggestions. One other sister also experienced favoritism, but the rest of my siblings had quite a time getting what they wanted from our mother and father.

I also learned that the love showered upon me by my parents was conditional love, although I did not give it that label when I was a child. My "most favored" status carried the price of expectations, whether good grades or a bubbly smile. This clearly came through to me on my sister's 13th birthday. Even though it was her birthday, she received no birthday present from our parents. However, on this very same day, I received a gift as a reward for the good grades I had earned at school.

Although I was only seven years old when this injustice occurred, I was very much aware that this did not feel good. Through this and similar experiences, I began to sense a separation from my brothers and sisters. Although I could do almost no wrong in my mother's eyes, as I became aware of inequities, her preferential treatment made me very uncomfortable. As a result, I changed my behavior and stopped mentioning things that I wanted or liked to my mother. Inside, I vowed to myself that if I ever had children, I would love them all equally.

Later, in my adult years, I understood the injustice as well as the blessing I had received from my parents. Even as a teenager I felt a lot of gratitude, although I promised myself I would do it differently when it became my turn to be a parent. The opportunity came sooner rather than later.

To Montreal

I decided I wanted to study to become an x-ray technician, and in order to get the proper training I had to leave my home town of Alma, population 15,000, and move to Montreal, Quebec. Once I proposed this idea to my mother, she readily agreed that I should go. After all, I had always gotten anything I wanted, and this was no exception.

It was in Montreal that I met my first husband. I had never had a boyfriend before, so I was pretty unsophisticated in the world of romance. Basically, I fell in love with the first guy who came along and decided I wanted to get married. My heart went out to this young man because he had suffered a great deal of rejection. This was because his father drank too much and his mother lived in a continual state of fear. She was the frequent object of her drunken husband's violence. I was extremely naive and thought that I could give him some of the happiness I had. I vowed to myself that I would love him and take away his pain. I noticed that even at the tender age of 16 (I was 18), he was drinking a lot. But I was certain that when we were together he would change.

In 1960 we decided to get married for a very good reason: I was pregnant. We did not tell our families about my condition. I just went home one day and said that I wanted to get married. My surprised mother said okay, but to whom? So we were married and returned to Montreal to live.

Pregnant in Peru

We were both Roman Catholic and my husband was involved with a church youth group. Even though I was studying to be an x-ray technician, the congregation of this church offered to send us to Peru as missionaries. Our assignment was to show the Indians in the Amazon that they should get married as we had and that the men should have only one wife. So off we went to Peru. Although I was pregnant, I had no real anxiety. I felt that everything would be fine. While in Peru we stayed with a group of Catholic Brothers, and we lived in a little house at the side of a monastery. The plan was for us to leave there and go to the Amazon region, but this never came to pass. We returned home after only three months, because the Monseigneur in charge at the time was worried that there was no doctor in the remote Amazon region of our missionary assignment. He did not feel it would be safe for me to give birth without benefit of a doctor. So when I was eight months pregnant, we returned to Montreal.

This short, three-month stay in Peru made an indelible impression on me. It was through this experience that many questions about religion began to surface. I had observed the Brothers at the monastery, and the way they treated the poor people who came to get food. I felt they treated people with disrespect and fed them like you might feed animals—uncaringly dumping food into tin cans. The contradiction between how it was *supposed* to be and how it *really* was struck me strongly. My attitude started to shift. I began to see the Catholic Church and various other religions as being institutions of control rather than of love. I also began to think about the position most religions take: our way is the *only* way. This struck me as a most decidedly ungenerous and unloving attitude.

My disillusion with the church was gradual, but the seeds of it began during our trip to Peru. I stopped going to church regularly, and eventually stopped practicing the Roman Catholic religion altogether.

Christine's "Supermom"

In the meantime, I gave birth to our first baby, a sweet little girl we named Christine. Unfortunately, our marriage was extremely chaotic and my husband was drinking heavily. He was also unemployed, so money was extremely sparse. I was going to school, and to earn some money, I began to make flowers from wire and nylon and sold them door to door.

When I passed the test to be a certified x-ray technician, I found a job. This paid much more than selling flowers! I worked part-time four hours a day. I was still breast-feeding Christine, so I was running from home to work and back. Because my husband could not be responsible to take care of the baby, I had to take her to day care. He worked off and on as a plumber, but when he did get paid, he drank away his paycheck. The home front was pretty rugged, but the hospital, as it turned out, was little better.

I was very enthusiastic at first. I wanted to help people and be part of a healing environment. But I learned pretty quickly that the hospital environment was not really a place for healing. I discovered the doctors often were not even examining the patients who came into the emergency room, but were prescribing for them anyway. I was very naive and remained that way for a long time, yet I began to question and ask, "Why do you order a head x-ray, if the patient is complaining of stomach trouble?" The Canadian government ran the health care system and people came to the hospital for regular checkups. It was common for people to have x-rays ordered by the doctors, even though they had no apparent problems.

I knew x-rays were dangerous; that had been taught to me when I studied to be a technician. I hated to see patients who came in for routine checkups being subjected to dangerous x-rays, when they were utterly unnecessary. I began to try to alert patients a little bit. Finally, I ended up working more nights and weekends because people needing routine checkups came in during the day, and at least at night and on weekends, people with genuine emergencies came in. Many of these people with medical emergencies really needed the x-ray procedures, so I felt better about it.

Meanwhile, life was very busy. I continued to work outside the home to support the family and I was taking care of my infant daughter. I tried to understand my husband and to take care of the home. I was experiencing a very different life in nearly every sense than I had enjoyed while growing up.

Confronting Fear

My father drank some, but when he did he became sweeter and never caused any problems. I thought mother was being overly dramatic to complain about this. But when my husband drank, he often disappeared. The most difficult thing for me to deal with was when my husband didn't return home for dinner in the evenings and I had no idea where he was. I would get terribly worried that he had become drunk and had had an accident, killing himself – or worse, killing others.

After countless nights of worrying myself sick, I finally did something which freed me from this terrible fear. I decided to confront it. I did this by mentally living through the whole worst-case scenario that I could imagine. I visualized my husband being drunk, having an accident and dying. I imagined the sadness that I would feel at being a widow. I went through the whole imaginary episode, feeling and seeing the different elements, one by one. Naturally, this had a tremendous emotional impact on me and I cried and sobbed as I visualized myself as a widow with my daughter, continuing to live my life alone, managing by myself.

This visualization process freed me. I discovered that by going through this process step by step and really confronting my fear, I could remove the fear. In fact, I discovered that it would not be that bad, because I was already living as if I were alone. I was taking responsibility for everything. After this, I realized that the cause of my worry was that I was afraid of losing my husband. I was very attached to him. By going through this process, I released him and no longer saw him as "mine." I realized that it was very helpful to live in the here and now. I thought to myself: okay, he is here with me now—and I stopped imagining a future time when he would no longer be there.

Through this simple process that I stumbled onto "by accident," I reached the point where I could say, "If he comes home, he comes home, and if he doesn't and he's dead—well, I already went through that in my imagination and I can live through that, too." So, when my husband returned home the following morning, I was very surprised and relieved. But better than that, I saw him as a gift. I discovered he didn't belong to me anyway. When he chose to be with me, that was great. I was content. My fear was gone because I had already experienced the pain and grieving in my mind, and felt all the feelings.

Detachment Won

I continue to use this process in my life. It has helped me learn detachment. Whenever a fear of losing somebody or something surfaces, I go through the same visualization process step by step. I think and feel the loss as if it has already happened. I release the situation and acknowledge my gratitude for the service that this person, thing, or circumstance has given me. I really believe this process is what has kept me happy inside, despite all the challenges I have created in my life.

My husband and I remained married for 13 years. My major problem in choosing to marry him was that I truly believed that I could change him. Through my experiences, I learned that was a naive and foolish approach to take!

Over-Radiated Robert

Soon after I stopped breast feeding my daughter Christine, I became pregnant again. I was working in a clinic as an x-ray technician, and had been confident the department of radiology where I was working was sufficiently protected against the dangers of the x-rays, with appropriate lead shielding. I was horribly wrong.

On my last day of work, before I quit to have my second baby, I decided I was too tired to take the x-ray plate I had just exposed into the darkroom. I temporarily put the exposed plate on the floor while I took another x-ray of the patient. What a shock I received later when I saw my own leg exposed on the plate I had temporarily put on the floor! My mind reeled at the realization that my soon-to-be-born baby and I had been exposed repeatedly. I tallied up all the radiation my baby, myself and the others who worked in that place had received. It was staggering!

I gave birth to my son Robert a few days later. I really didn't know what to expect. I was aware that radiation is even more dangerous when cells are multiplying rapidly—as they certainly are when a baby is growing during pregnancy. All I could do was hope that everything would be fine.

Robert's birth was not an easy one, because he arrived feet first. He was nearly folded in half. Problems were apparent from the beginning. When I tried to breast feed him, he had difficulty breathing and drinking at the same time. Nonetheless, I persisted for three months, knowing that this was the best way to give a child a good start in life. But I could not continue beyond three months. In addition, Robert contracted a continuous stream of ear infections and nose, throat and eye problems.

Life at home continued to be chaotic and unpredictable. I ran from the hospital to the police station to home, and then made the circuit again. My husband was arrested for driving while inebriated. He had a hard time assuming responsibility for his own life, and he certainly wasn't ready to assume responsibility for a family.

My son Robert was constantly sick and by the age of six he had already undergone four surgeries. It was at this age that his glandular system completely broke down and stopped functioning. As a result, Robert's body was unable to produce the critical growth hormone to mature—he simply stopped growing. All of this was very tough for him to handle psychologically. He could not participate in children's games and children could not relate to him because of his size and physical weakness. Even today, in his thirties, he could pass for an 18-year-old.

Robert's responses to his environment and surroundings were not adequate or appropriate. Due to the lack of hormones, he had no reflexes to run or take flight. Many times he had to be pulled back from dangerous situations, because he simply was unable to gauge the speed of objects or people. During his teens, it became even more difficult for him as he refused to take the medications necessary to sustain his life. He suffered tremendous psychological distress.

When Robert was only two years old, I became pregnant again and gave birth to another son, Francois.

Church Dogma Again Questioned

Up to this point, I had followed the dictates of my church and had used no contraceptives. This, coupled with the fact that I had an irregular menstrual cycle, made it impossible to use the rhythm method effectively. I therefore decided that I'd had enough of the church's control in my life, and I had a tubal ligation to ensure there would be no further pregnancies.

To take such a radical measure was a big step for me, because my family was still Roman Catholic; in fact they are to this day. But once I took a stand for what I believed, I rejected the whole Roman Catholic dogma. At the time, I was simply unable to discern the true from the false. The only thing I knew for sure was that this religion didn't make any sense to me, and I ended up following my heart.

Ethics Versus Economics

When I was growing up, I obeyed my parents and my teachers. I learned that they treated me very well because I obeyed them. The more I was exposed to the world, I realized that this system worked for a child while growing up, but it was not appropriate or helpful for a full-grown adult. It began to dawn on me that the established systems—educational, medical, and political—were not really geared to help, but instead took advantage of those who could not and would not rebel.

At the hospitals where I worked, it seemed that countless tests were prescribed, to protect the doctors and/or to generate money, but they did not really help the patients. Of course, when a patient has no basic knowledge of how the body functions, how are they to judge? When they have a pain they go to see the expert, the person who is supposed to know. They trust their doctor. But what happens if this doctor was taught and trained by a group whose primary goal was the accumulation of money and power? The students, in this case the doctors, simply applied what they had been taught. I was deeply disappointed with the medical establishment in Canada, after observing and working in hospitals for twenty years. I witnessed an entrenched pattern of covering up and suppressing symptoms, using drugs, radiation and surgery.

Leaving the Establishment

Today, I have great respect for those physicians who do reach outside the established medical system. Often these courageous healers place their right to practice medicine in jeopardy, simply because they know there are many natural, non-invasive ways to help the body heal. They also know that the body is only one aspect of a person, and that the mental, emotional and spiritual aspects must also be included if the healing is to be genuine. Otherwise, a problem may be temporarily resolved, but unless an idea, belief or emotion is changed, another will crop up in its place.

In the early 1960's I didn't know what else to do but criticize the system. I talked to patients whenever possible. I tried my best to protect them with lead aprons and other tools to combat the effects of radiation. I even kept questioning the doctors if I felt the prescribed x-rays were unnecessary. It is only since moving to the United States in 1984 that I have seen the good being done by some doctors. But back in the sixties, I knew something was not working the way I had envisioned that healing should be. I now can say that the government ran the medical system in Canada, so there was no incentive to be healers instead of prescribers. The system did not reward healing, only the generating of income.

Challenging My Golden Goose

At the same time that I was railing against the medical system, I was also benefiting by its services for my son. Robert was exceedingly fortunate to be able to receive the extract of growth hormone, to compensate for his pituitary's failure to produce it. This growth hormone was very rare and very few children could receive it.

I was in tremendous conflict between the genuine need for help for Robert and the failure of the existing medical system. Compartmentalization was rampant. There was a specialist for each little part of the body, so that each one knew little of the overall situation with their patient.

For a while, I worked within the established systems, hoping to change them. I went through three phases. My first phase was to try to change the medical system at the hospital level. When that didn't work, I went on to phase two. I joined the Separatist Movement and tried to correct things through political means. I quickly discovered that was even worse! Finally, as a result of a three-year university experience studying for a degree in psychology, and unable to tolerate the educational system, I tried to reform that too. But I discovered that I could neither stand it nor change it. So I left.

At the time, I didn't realize that all change must come from inside ourselves. Today I know that each individual has the sacred responsibility for their own life. When we give our power away to a partner, religion, doctor, government—whatever—we grossly negate who we really are, a Child of God. As long as we believe that others are responsible for us and for our well-being, we cannot do a thing. It is calamitous for us to believe that a person or group should make decisions for us that will lead to our happiness. This illustrates an appalling lack of self-knowledge. If we abdicate self-responsibility in this way, then we don't even know ourselves enough to make a decision that can make us happy.

Victory Over Fear

Chaos as Catalyst

Life at home improved, at least financially. My husband no longer worked intermittently as a plumber. After we had been married for five years, he studied to become an architectural draftsman and eventually took a full time job. Although he still drank heavily, he was able to maintain employment.

By 1972, after twelve years of marriage, we had reached the point where we owned a house and had more financial stability. Nonetheless, alcohol addiction continued to traumatize our lives. At that time, I had never heard of such a thing as co-dependency, and I did the best I knew how to do. I wanted so much to hide my husband's drinking problems from the children and spare them this anguish that I would send them to bed very early, thus avoiding his heaviest drinking hours at night. However, by the time Christine was eleven years old, it became pretty difficult to pack them off to bed at 7 p.m.! Things were starting to unravel, but I didn't know it yet.

One afternoon after work I decided to wash the outside windows. I found a ladder and propped it up against the side of the house. I was forced to place it at a somewhat precarious angle in order to reach a particular window, so I asked my sons to sit at the bottom of the ladder to steady it. Not surprisingly, the children, tired of their job as ballast, wanted to run off and play. I agreed and let them go, but at the first wipe of the window, the ladder and I went crashing to the ground. The fall was quite severe, and after being taken to the hospital it was determined that I had broken my clavicle, and the trauma had also dislocated my shoulder.

What a pain. . .what a blessing! Richard Bach's *Illusions* so aptly says, "There is no such thing as a problem without a gift for you in its hands. You seek problems because you need their gifts." Well, this "accident" was a living example of what he wrote about.

My accident caused me a great deal of physical pain and distress. It also made it impossible for me to go to work. This was the first time since we had been married that I was unable to work. Suddenly. . .*finally*. . .I had time to think. I had time to evaluate and review what was happening with my life.

Body Immobilized - Time to Evaluate

At first when I was immobilized and forced to stay home, I felt guilty for not going out and working. It turned out that one of my lessons in this whole ordeal was for me to learn to do nothing. This was most difficult for me, but very, very important.

It began to dawn on me that my super-busy life had been serving two purposes. First of all, it kept me too busy to think. And somewhere I had decided that the less time I had to think about unpleasant aspects of my life, the better. In my head, I had made a commitment that I was married for life. Key words in my life were "shoulds," "musts," and "have to's." Therefore, they translated into my own private orders to myself: "Nicole, you must stay married, you should tolerate your husband's drinking, and you have to hide his drinking from the kids." Simple, direct and ironclad. Until my so-called "accident."

The second purpose of my very busy life was the pride I took in being able to do two or three things at the same time, have a thousand interests, and make everything from scratch. I always enjoyed having nice things and in the beginning when money was scarce, I sewed clothes for us out of necessity. Later, I began weaving fabric for drapes and bedspreads. I made furniture, gardened, painted and created pottery. You name it, I did it. I kept myself totally busy throughout my waking hours. And all the while I worked full-time outside the home, and raised three children with no help from my husband. On top of all that, being married to my husband was almost like a second job. I felt that I not only had to hide everything from the kids, but also maintain a good outward appearance. I hid the fact my husband had a drinking problem from everyone. I was running from everything.

Eye Opener

The six months it took for me to completely recover from my shoulder injuries provided me with a real eye opener. I discovered I was not only trying to hide the situation from my kids, friends, and family, but also from *myself.* I had refused to admit that basing a relationship on the premise that "I will change him" simply did not work. I began to recognize that this was not unconditional love, it was very conditional, very flawed human love. The implied deal was "I will do this for you and you will be nice." The fracture of my clavicle, which I unconsciously chose to experience, was a painful way to come face to face with this truth, but I needed it to happen this way in order to face the facts.

Working in a hospital emergency room, I repeatedly saw people with so-called accidents, and it was as if each person had no role or responsibility in creating their accident. I have learned that regardless of what happens, we create everything in our lives—including "accidents"—whether or not we are aware of it at the conscious level.

The therapeutic physician who helped me to recover physically asked me many questions that encouraged me to open my eyes. I suddenly realized that I was in control of what I was doing in my life. I had lost myself in all the different roles I had played; but these were not truly who I was. I discovered there is a big difference between being and doing. I had certainly been an expert *doer,* but I had managed to skip the *being* part.

Thinking With My Heart

As a result of being totally self-sufficient, managing all the major family decisions, raising the children, and being the wage earner, I had developed the masculine qualities of myself to an extreme. I controlled the physical matter in my world through doing, logic and brain work. Now I decided it was time to listen to my heart and to let my feminine side blossom. I needed to shift from "loving with my brain" to "thinking with my heart."

I had learned at a young age that I could succeed by not talking about what I wanted. It worked well to keep my parents' obvious favoritism toward me hidden and less obvious, as they could be counted on to provide my wants and needs, nonetheless.

When I was the mother, I was the one providing for the family's needs. There was nobody thinking about my needs and fulfilling them. I knew exactly what my husband liked and made it my business to do everything to please him. I had placed him first, my children came second, then came my work and other things followed. I had not even included myself on this list. I was living my life through others. But this is what I thought love and life were all about. "C'est la vie," as we say in French.

My Charade Ends

Finally, I reached the decision that I wanted to find myself—*who I really was*—and to give my children a better life. The charade was over!

I had become extremely concerned that unless I did something, the situation would only get worse and worse. One day my husband came home drunk; Robert was ten years old and could see that his Dad was in an inebriated state. Robert told his father that he was acting foolishly and that he should sit down. Robert took a very strong tone with him in telling him to "sit down and be quiet." He was almost reprimanding his father. Naturally, his father resisted the orders issued by a ten year old! It was then that I realized that in a couple of years there could be physical fights between the two of them.

It was this particular incident that served as the catalyst for my decision. I did not want my children to physically fight with their father. My husband had fought with his own father and I could see that this was going to be played out in our home too, unless I did something.

The discipline my husband maintained in our household was very firm. The children obeyed him out of fear and the threats of violence. And although he did not beat them, he could be verbally and emotionally abusive, and sometimes he threw things. So I consulted with the children and together we decided that divorce would be the best way to go.

The children jumped at the chance to have more freedom and to be released from this oppressive lifestyle. I tried to make them understand that I truly wanted them to continue to have a relationship with their father. Yet I reasoned that some breathing room would make this more of a possibility.

I can honestly say the kids were thrilled with our decision, and so was I. They were perhaps even more thrilled than I. After all, they had not been able to invite friends to come to the house for fear their father would show up drunk, and they had submitted to a rigid, strong discipline, almost like being in the army.

My husband's intentions were good, and there had also been good times with the children and they all loved each other. But the situation was going from bad to worse and I had to stop the negative escalation.

Fear of Death

When I asked my husband for a divorce, he told me that he would kill me if I went ahead with it. I had no trouble believing he would follow through on this threat, since he became violent when he drank. So for six months I was afraid of divorcing him because I was afraid of dying. I had conquered many fears in my life, but the fear of death was not one of them. I didn't know how to get over this.

Then one afternoon, coming back from doing an errand after work, and wondering how I could get out of the marriage alive, I had the most profound experience of my life.

My Near-Death Experience

I had stopped my car behind a garbage disposal truck on a narrow road that ran along the river leading to my home. The driver signalled me to come around and pass him. I realized there was a turn in the road not too far ahead, but in my naivete, I followed his signal and proceeded to pass. Much to my horror, as I swung into the oncoming traffic lane, I saw a big truck full of gravel coming toward me very fast. There was no way I could avoid a collision. A moment before impact I said to myself, "Well, this is it! I'm out of here!" And with that, my awareness lifted out of my body.

I did not have a clue about any of this kind of thing. Today we frequently call these "near-death experiences" (NDE's). I wasn't aware of such a possibility. All I know is that when I saw the truck, I knew there was no way for me to be safe while remaining in my body. I had already left my body when the truck hit my car. I discovered that when you are in a situation like this you can decide to go, and it's very simple to accomplish. Despite the fact that I had no conscious understanding or knowledge, I simply said, "This is it, I'm gone," and that was that. Basically, I didn't even see the accident. It was kind of nice.

During my years working as an x-ray technician, I had seen thousands of people suffering from automobile crash injuries, and I did not want to experience the kind of physical pain I had witnessed. I figured I had enough emotional pain. I certainly did not need to add physical pain.

Of course, I had no idea what to expect. I was pleasantly surprised to find myself floating in a body of Light going through a tunnel. The significant events of my entire life passed in review. I remember the criteria that I used for judging all my experiences was how much love I had given. I was the only one who sat in judgement of myself.

I was shown my own criticism of my mother for the "injustice" she committed toward my brothers and sisters, because she didn't appear to love them as much as she loved me. Much to my surprise, I found out that my love for my children was equally conditional. I realized that I had loved with my brain, and that I required a reason to love. I knew I was not bestowing unconditional love on my children. I was aware that my intentions were good, even if I was far from having demonstrated unconditional love. I did my best.

After this, I continued towards what I thought was heaven. As I approached the end of the tunnel, I became aware of balls of Light on each side of me. I felt good about them and began communicating with them via telepathic impression. They told me that it was all right if I proceeded, but they asked "What will happen to your three children?" As soon as the balls of Light mentioned my children, I told them to "hold on" and I would come back. I decided I should go and check out my body to determine my injuries.

Re-Entry

Suddenly I was back in the car, hovering near the roof. The car was squeezed between the truck and the tree, and resembled an accordion. The steering wheel was touching the driver's seat and my body was on the passenger's seat. I had not been wearing the seat belt. I inspected the body fairly easily since it was summer and the body was dressed in a top and shorts.

I was stunned to observe that a virtual miracle had happened! There was not a scratch on the body! Wow! Aside from having squished the flowers that I had just bought by sitting on top of them, everything appeared intact. Thus reassured, I decided to re-enter my body. Once I re-joined my body I noted there was no pain, nothing.

With great excitement, I opened the car door and got out. The three men who had witnessed the entire accident could not understand why I was so excited. They asked me a couple of times if I was all right and I replied, "I have nothing wrong, I have nothing wrong!" as I touched my body to make certain. I was in a state of euphoria!

When the men offered to help me, I replied that "you have helped me enough," thinking of what I had just experienced—my wonderful NDE! I then decided to walk to my home nearby.

Instant Freedom

Two hours later my husband arrived home from work. Having seen our totalled car on the road, he didn't know what to expect! I was in the kitchen fixing dinner and thinking about the powerful experience I had just had, when he opened the door and asked me what happened. I replied all in one breath, "A big truck hit me and it didn't kill me, so you can't kill me either and I want my divorce right away!" My near-death experience had freed me from the fear of death! My husband had no more power over me!

I got my divorce six months later, thanks to this life changing experience! The children stayed with me and I continued to work to support them, since I received only a very small support payment. My life was full and exciting. I had a lot of energy because I had lost the fear during my near-death experience. Prior to this, my fear had consumed so much of my energy. It was an exhilarating feeling. *I* was in power, doing what *I* wanted to do.

Recognizing My Own Responsibility

After I realized that I had given my power away, I could not in any way blame others, including my ex-husband! The oppression that I experienced in my marriage didn't really come from my husband. He had done the best he knew how. The problem stemmed from my not speaking my truth about how or what I wanted out of a marriage. Instead, I had allowed myself to become enduring, compromising, believing that this was the way that a woman should be. During the divorce process this truth dawned on me, and my husband and I made a serious attempt at reconciliation. But it didn't work. There were simply too many holes in the fabric of our relationship and there was no way to repair it.

At that time, there were many things I didn't understand about my marriage. As a way to help me clearly see the situation, I began writing everything that came to my mind. It was tempting to put all the blame on my husband's alcohol problem, but this was not really the answer.

Putting all the issues down on paper allowed me to become

more objective and see my equal responsibility in creating the challenging situation of our marriage. For my part, I knew clearly that I had learned my lessons and I would never deny my own values and integrity again. By taking responsibility for my own part in this whole marriage experience, I was able to change my thinking and release the old beliefs that had kept me in that relationship. More importantly, I was able to forgive myself for having accepted such a role in my marriage.

Prior to this, I had always felt responsible for others, and thought that others were responsible for my well-being and happiness. But after my NDE, I realized that I was the only one who could make changes in my life, and only I was responsible for meeting and fulfilling my needs. I discovered this was true for others as well.

I gradually became aware that the only way to change the world was to change myself! What a relief to remove from my shoulders the heavy burden of feeling that I had to change the world. In addition, I realized that any inner fears an individual has not only creates a separation, but also takes away one's power.

A Tool for Personal Clarity

During the process of writing down everything that came to mind, I gradually learned to detach from the situation and become objective. With this new-found objectivity, I was then able to map out my future plans. I wrote down my goals, hopes, dreams, wishes, etc. and then sorted them out into a life goal for the next period of my life. It was a great success.

I have since refined this process, which was so helpful to me after my divorce, and now I use it regularly. I encourage everyone to try it. Napoleon Hill's book *Think and Grow Rich* points out that every wealthy man he interviewed WROTE DOWN their goals, and they were extremely successful (not just financially).

This was also true for me. Do not just think and dream, but write down your plans and goals, then review them and update them regularly. You will be well rewarded.

Rebuilding My Life

Striking Out Anew

Our divorce was final in 1973. I felt a great sense of freedom, and as always, a sense of adventure. I continued to work part-time as an x-ray technician, but on a greatly reduced schedule. I went back to study Human Relations at the University for three years, but I didn't enjoy the educational system, so eventually left. After that I got into a new line of work that I loved and it absorbed most of my time.

My new venture began when I was on vacation. While I was away, I visited a place where they had different rocks and agates. I fell in love with the crystals I saw and wanted to bring these beautiful pieces home.

At the time, I had absolutely no idea that crystals had any particular powers or metaphysical purposes, I just knew that I loved them. I learned how to cut the crystals, took classes in polishing, and purchased a machine to polish them. I did that because I loved to touch them and I loved the rainbow colors they refracted.

Crystal Tycoon

I ended up having so many crystals that I didn't know what to do with them all, so finally I started making jewelry and selling it. This was especially surprising since I had little interest in jewelry and wore almost none myself. None, that is, until my crystals came along. I enjoyed wearing crystal jewelry and, happily enough, so did many others.

I began selling the jewelry that I made from crystals, and in a short time I had a very lucrative business. I had twelve distributors selling my jewelry all across Quebec. I was making so much money I didn't have time to spend it. People loved wearing the jewelry and I loved making it! My crystal jewelry business flourished for seven years, and in the meantime, I continued to work two days a month as an x-ray technician in the hospital emergency room.

Family Tragedy

My daughter experienced an extreme tragedy during this time—she was raped. It was horribly traumatic for her, and she struggled with a variety of ways to try to come to grips with what had happened. She went to a psychologist for a year, and tried everything she could think of to help re-establish her equilibrium and eliminate her fears. Fear was always present with her—she could not sleep alone, she was afraid of the dark and did not want to go outside. She was extremely nervous and very depressed.

I so much wanted to help allay her fears, but nothing that we tried helped. Then one day a friend of hers introduced us to a group that used a mental technique for dealing with the trauma. I observed my daughter go through this mental process, and was astounded that within half an hour, the whole trauma from the rape was eliminated. I was completely in heaven as I observed this seeming miracle.

I noticed that this mental system seemed to be very powerful. So, after a family discussion, my children and I agreed that we wanted to go to the United States and learn how to do this. It had made that great an impact on us!

Destination USA

In less than two weeks time, I sold my crystal jewelry business, my home and all my material possessions, and with only the items held in two suitcases, left Canada in order to learn this method. My house and business sold rapidly because I put such a low price on them. When people learned the price I was asking for my house, I had six offers the very first day!

I have deliberately not identified the group with which we became involved, because this turned out to be one of those "learning experiences." Six months after we arrived in the USA, my son Robert became extremely sick, and I had to attend to his needs. This turned out to be a blessing, as it provided my escape from this disturbing situation. So I left the group and rented an apartment in Los Angeles in order to take care of Robert.

This necessary departure allowed me to see my truth regarding some organizations. It was a deeply impacting experience, a clear demonstration of how people in authority often abuse their power, controlling people with their teachings and programs. It is important to acknowledge that they usually have some effective techniques. My discernment lesson became the separation of the good things from the undesirable issues. This can be a difficult lesson for some people to learn, as they have already given away their personal power without realizing it. This is one of the major lessons for human beings through all time, because disempowering ourselves means that we lose sight of who we truly are—"spiritual beings"—and reclaiming ourselves becomes the first and most important step that we take.

Exploring Alternatives

Nonetheless, the system that helped Christine rid herself of the fear and trauma from her rape gave me my first inkling of the power of the mind. This whole arena fascinated me, and I wanted to know more. And so I began many years of study and discoveries. I also wanted to explore alternative medical therapies that might possibly be helpful for Robert, so I researched various energy modalities. I discovered, for example, that magnets, radionics, the Rife frequencies generator, and homeopathic medicine could each be used to make positive changes in the body. I also learned about the healing properties of crystals and began to use them again.

I had always been aware of healing energies in my hands, which I used mainly with Robert. I took some attunement classes at that time, and studied *Reiki,* which is a Japanese form of hands-on healing, later becoming a Reiki Master. During all these years of research, training and investigations, I became involved with different companies and sold their equipment and products.

Partnership Re-Explored

It had been sixteen years since my divorce and I had lived a good number of years as a single woman. This had been an eventful, adventure-filled time, but now I decided that I was ready to find a special person that I could love and be loved by. So I set my intention to find a man to be my partner. By now I had realized many, if not all, of my dreams, and even the fantasies of being on my own. Now was the time, I felt, to re-experience living in a relationship.

A friend of mine had good luck placing a personal ad in the classifieds and I decided I would do the same. I wrote my ad to put in the Los Angeles paper, but never placed it. The day after I had written it, I picked up the LA Times and discovered a personal ad that was almost the same as the one I had written. So I responded to his ad, he called me and we met the following week. Three months later we were married. Our three months of dating were wonderful. We had a lot of pleasurable times together including traveling, dining, and love-making. Everything appeared to be perfect.

However, a very short time after our marriage, I realized that my expectations regarding our marital relationship would not be fulfilled unless I was willing to make a great deal of effort. I recognized I could make these efforts, as I had done in my first marriage, but after evaluating the importance of this, I decided to let go of my expectations and concentrate on my spiritual growth.

Our marriage quickly settled into a pattern, and became quite unremarkable We made love once a week, went out for lunch or dinner daily, and each did our own thing. We also did some things together, but the excitement present during our three months of dating was gone, and there was little romance to speak of. Nonetheless, this became agreeable with me as I had already decided not to expect too much. I now wanted to focus on my spiritual growth because I had served my "little self" for 49 years, and felt it was time to serve my "Higher Self."

Surrender

Surrender became my next goal. I realized that I could have asked my husband Rene for what I wanted in this relationship, and with effort, I probably would have received it. But I remembered my first marriage, when I had wanted to change my husband. I knew that this had not worked in a satisfying way, and it was foolish to think it would work well now. So I decided just to love Rene the way he was. I reasoned that since I was already happy when I got married, I didn't *need* him to make me happy. I did think however, that I could expand on this base of happiness, and continue to grow with him.

We became interested in an organization called *Eckankar,* a spiritual study of sound and light. I was reading their books and meditating every day and focusing on my spiritual life.

Music in My Head

About six months after I made the decision to surrender and to serve my Higher Self, I received a big surprise. Rene and I had come home from lunch. It was July and quite hot, so naturally the air conditioning was running. Suddenly, out of nowhere, I began to hear music. I didn't know where it was coming from, but I reasoned it had to be coming from somewhere in the house. I shut off the air conditioning, the refrigerator, and checked every corner in the house, but could find no source for the music.

I was baffled, but I gradually realized that the music was not in the house, but inside my head. And so the internal melody began—

and continues to play to this very day. I heard it only briefly at first, then gradually for longer periods of time, and within a month the music was almost constantly present. The music is gentle and very soft. It sounds like a distant chorus of people, and it is a melody that continually repeats itself. I used to wonder if I'd grow tired of hearing it constantly, but I haven't. The more I hear it, the more I like it.

I noticed that when there was a humming sound nearby, like the refrigerator motor for example, it seemed to amplify the music and make it louder. I also noticed that when I got overly excited, or was mostly using my left brain to think or process, I was unable to hear the music. For example, I decided to visit my family in Canada after many years of being unable to see them. When I went to Canada I had been hearing the internal music for several months. It stayed with me on the plane, but once there I didn't pay any further attention to it as I was so excited to see my loved ones. I became so excited that I no longer heard the music. If I get off center, I cannot hear it.

It was five weeks after I returned home before the music began again. After my return I tried everything to bring the music back, including singing the tune inside my head. I was very upset with myself. Finally I surrendered. One day when I was not expecting it and was making no effort, to my delight, the music returned! This was when I decided to try to remain in balance and not get overly excited. I have since discovered that this music serves as a barometer regarding my state of balance. When I am in balance, I hear the music. So I learned that extremes do not work well for me.

My husband and friends were used to hearing me sing, or listen to classical music and opera. But this all changed once the internal music began. Hearing this melody puts me into such a blissful state, that I want to remain there as much as possible. No music in the world is more desirable.

I knew from the beginning that this internal melody was a spiritual gift, and I loved it. I still love it! At first I thought this was a gift solely for me, much as other people have their own spiritual gifts. But I have since learned that this is not the case, that the "music of the spheres" is everybody's music, not just mine.

Re-Prioritizing

When the music began, I noticed I was changing. Effortlessly, I was detaching from physical and material desires. Detachment became automatic because I had something far superior—a deep, inner peacefulness that came from hearing the wonderful internal music.

I developed an indifference toward worldly things. When I had material possessions that was fine, and when I didn't have them, it was still fine. I truly came to know more than ever that happiness resides within. What comes from outside of ourselves can certainly bring excitement for a time, but it also fades away. I discovered that when we look for pleasure in "things," in material possessions, we always need more.

I also noted when the person who "makes us happy" is external to ourselves, like a partner, a child or an associate, we try to control them. We want to be certain that they will behave in ways that "make us happy," and we want to be sure that they continue to behave that way. However, we all know deep inside ourselves that we really can't control someone unless they permit it. And since they have the choice to change at any time, this brand of "happiness" is fickle indeed! Obviously it is quite fleeting, and as we inevitably lose our "happiness," we develop all kinds of fear-based behaviors.

It is a very different situation when partners in a relationship are not dependent upon each other, but are simply together because they want to be with the other person. In this ideal situation, there is no need for control. Each person is whole and complete; coming together doesn't make either one more complete.

If, however, the relationship is based on two incomplete people coming together to complement each other, a false sense of wholeness is experienced. This approach is based on which fragments each individual brings to the relationship, and works to some degree as long as no changes or shifts are made. Problems arise when one of the partners wants to make a change in any area, and the other one doesn't want to go along with the change. Suddenly, problems arise because the relationship has been thrown out of balance.

From Pleasing Others to Pleasing Myself

In my own relationship, my detachment was made possible thanks to the beautiful internal music I was hearing. I developed an entirely new way of loving and being, learning to love Rene just the way he was. I no longer needed to play games to get what I wanted from him, and I no longer needed to give him what he wanted in order to get what I wanted in return.

This was a big shift for me since I was an expert at pleasing. I had developed it to almost an art form in my first marriage. But no more. Suddenly I realized that I just wanted to be me. I wanted to stop doing things that I thought were "right." Instead, I wanted to focus on being the way *I felt* was right. I shifted away from my point of view of loving with my head, to thinking with my heart. I was feeling on top of the world and thought that this relationship would be the one that could last forever. I had a surprise coming.

Relationships are great teachers, and mine was no exception. Ordinarily, I don't get angry easily and don't have a temper, but this was not true during my marriage with Rene. I was not a pleaser anymore and just wanted to "speak my truth," but there was something else. Rene was there to push emotional buttons I didn't even know existed.

I was pleased with my progress concerning detachment from "things." I've never been too concerned about the opinion of others, but I did care very much how my husband felt about me. We played out an unpleasant kind of relationship-dance. He'd say something hurtful, or something that I felt was not true, and I'd get angry and answer back. He would act "confused" and demand to know why I was getting so angry! It was like beating somebody and then asking them why they were acting as if they were hurt! I'd try to tell him that I was angry because of what he said to me, but it was a futile dance we were doing together.

My Sixty-Day Experiment

Then one day I read the book *The Non-Human Experience* by John Prince, and decided to experience the book's sixty-day program. I committed to: 1) surrender everything, become totally detached and impersonal to my material sense of existence, 2) focus on spirituality rather than materiality and all action that I took in the outer world would be Spirit directed, and 3) express loving gratitude for all the good in my life, each day upon awakening.

John Prince said in this book: *"If you will truly dedicate yourself to living as a spiritual being for the full sixty days, you just might wake up one morning and find that you are living in an Island of Light, where everything that you could possibly desire is already in perfect expression as form and experience. . . and Truth is your eternal shield."*

I decided to follow this program because I really didn't like getting angry with my husband. Whatever he said to me, regardless of how critical he was towards me, I would not let it touch me any more. I realized that if I was able to become completely detached from his viewpoint, then I would no longer become offended and not get entangled in the situation. The problem had been the "little me" or my ego.

When I began to experience my life as a spiritual being, Rene could not hurt me any more, since only the ego can be hurt. Within a month I reached the point where I no longer reacted to anything he said. He really didn't like this, because my neutrality and lack of desire to be "hooked" by his taunts, meant that he had lost control over me.

In retrospect, I realized that I got more angry in the three years I was married to Rene than I had been in all the other years of my life put together! This marriage was a blessing, even though three months after reading the book we divorced. It was a tremendous force in my spiritual growth. At this time, we both value our continued friendship.

Love at First "Listen"

Meanwhile, the internal music continued. My heart really didn't need anything else, but my mind wanted to understand where this music came from and why I was constantly hearing it. I began to research music and sound and I attended many seminars on the subjects.

One week, three different people approached me about Sound Therapy. Needless to say, this was synchronicity at its best! It was love at first "listen." I became utterly fascinated with sound therapy. I knew instantly that sound was my calling, and immediately got involved with it.

In January of 1992, I saw a video about Sherry Edwards and her sound work. In the 1970's, Sherry realized she could hear sounds unable to be heard by the normal human ear. Her rare ability to hear in this way, coupled with her perfect pitch, enabled her to produce equally rare sounds. She could listen to a person speaking and instantly "analyze" their voice. A voice in complete balance will include all notes in an octave. But few of us have this total balance without doing some sound therapy. She could hear or identify the "missing" sounds, and produce them for her clients. Sherry was an early pioneer in sound and brought it to the scientific community. I respect her a great deal for the work she does.

Once I became enthralled with sound, I began to experiment on myself. I played around with different approaches that I had heard or read about. I also used information that I had learned of while I had studied and worked with different energy modalities. I based my experimentations on some simple truths:

Truth #1. Everything and everyone in the Universe has a certain ideal vibration.
Truth #2. Resonance is the frequency at which a person or an object vibrates.

Sound waves produce resonance. In fact, all things both animate and inanimate vibrate at a certain frequency. I noted that each one of us resonates with certain people, ideas, things and places, and others we have no resonance with whatsoever. But where was I to begin, and what exactly was the appropriate starting point? It was a big quandary.

In The Beginning was The Word

I was pondering where the starting point was, when suddenly it came to me like a flash out of the blue. "In the beginning was the

Word." Of course—the "word" was sound! This famous quote, from Genesis in the Bible, meant that in the beginning there was sound. It was then that I recognized that **sound is the key to everything** here on earth.

I realized that when we look at the full spectrum of frequencies—matter, sound, light and the various rays like x-rays, gamma rays, etc.—that matter is the most dense and the lowest frequency. As we move up the frequency spectrum from matter to sound to light to colors to ultraviolet and then other radiation, sound remains the closest neighbor to matter. Maybe that is why it is so powerful. Matter responds very well to sound because it is the closest neighbor in the frequency spectrum.

I believe that sound, or "The Word," is what transforms us from beings of Light into dense matter when we come to earth. I also believe that we can return to spirit and our true state as a being of Light, by "riding sound" or following sound, to take us there. Sound opens doorways to other realms and dramatically expands our horizons. Sound bypasses the logical mind, which is the source of any resistance. Sound is a right brain activity which the left brain doesn't understand.

Have you ever thought how interesting it is that the first three letters of Earth are E-A-R? Is it that sound creates Earth?

Increasing Awareness

As a result of my near-death experience, it became clear to me that I was Soul or Spirit with a physical, emotional and mental body. This knowingness was present, but I was not constantly aware of it and spent most of my time raising my children, working and fulfilling my own earthly desires. I had a very good intuitive sense that I followed sometimes, but most of the time I used only my reasoning. I can say now that my reasoning got me in trouble, and my intuition helped me to get out of it. That is why I learned to follow my intuition more and more.

The Four Ways of Receiving Impressions

I learned that people receive impressions from their Soul in four different ways. Some people have their **inner ear** open, so they can hear the little voice telling them what is the best thing to do in certain situations. Initially, this happens when there is danger. I cannot remember hearing this little voice, only hearing my inner music. Others will have their **third eye** open, so they can see auras and images to guide them. I sometimes see auras, but I cannot say that my third eye is truly open. The next inner sense, that is also very powerful, is the **gut feeling**, located in the stomach (the third energy center, or chakra). It is another way of receiving information about whether or not to take certain actions. I was not very developed in that one either. The fourth way of receiving information, which is often the most difficult to acknowledge, is the **intuition.** It is an instant inner "knowing," which overrides reasoning, not allowing the brain time to analyze, especially in an emergency. Intuition is easy to deny, because it is unrelated to any of the five physical senses.

My Intuition Lessons

In 1983, I was walking on a busy shopping street in Montreal where some of the stores had big neon signs hanging above the sidewalk. Suddenly, for no apparent reason, I ran ten to twelve steps, just as a big sign crashed one foot behind me! I had hastened just enough to get out of the way. If I had taken the time to question my intuition, I would have been injured, or maybe dead. This was a significant intuitive message for me, and as a result of this experience, I continued to listen to my intuition in critical times or emergencies, and tried to use that wonderful gift in a more conscious way in my day to day life. It never fails me, and I have discovered that responding to it is always the right thing to do.

In normal life, my intuition becomes apparent as doubts about doing certain things, so I have learned to pay attention in this arena too. I remember in 1987 I was dating a university professor who impressed me because he was very knowledgeable, had a phenomenal memory, and was a minister. I saw him as a kind of superman, because when I learn something, I try to apply it in my life; I assumed that he was very good at doing this too. So we decided to become engaged.

I gradually began to have some doubts as I realized that his ego was huge, yet the heart connection was not there. Also, I was beginning to realize that he was not applying his vast amounts of knowledge as useful tools for his life. At this time, somebody loaned me a book about how to use the pendulum, and I discovered that it was easy for me to use. I began to play with this new "toy" and was impressed by the responses. The first evening I asked the pendulum some questions about my relationship, and all the answers confirmed my doubts. After a restless night, really recognizing and listening to the doubts that I'd been having, I decided to break the engagement the next day.

I realized that every time there was a little doubt, my intuition told me to "pay attention." The pendulum became a way to confirm my truth. At that point in time I needed to see something visible in order to prove that what I perceived was true.

People who can hear the little voice, see in the third eye, or have gut feelings, have this physical proof; but for all of us that receive primarily intuitively, it is often difficult to discern between what comes from the higher part of our Self, and what comes from the lower part of the mind. Our left brain, logical mind has been trained to control and make the decisions, instead of allowing our intuitive, right side of the brain more opportunities. We must re-train ourselves to use this skill, by honoring our intuitive hunches. Since my intuition is the Higher part of myself I declared it the *Master,* and my lower mind became a wonderful *servant.* It is important to note that the pendulum is a limited tool, and can be influenced by surrounding energies if we are not careful. I learned to contact my Higher Self before using the pendulum, so it became an extension of my Higher Self.

Assuming Responsibility

When I began working with sound, I became more attuned to finer energies. My physical body became more sensitive, and I could perceive much more readily and directly than ever before. I quickly realized that sound was very powerful and required great care and concern. I wanted to ensure that my sound work only helped people, and certainly did no harm.

I intuitively knew about the power of sound and had apprehensions about its use. Several months after I started doing my sound work, I met my first channeler. This person introduced me to specific guides, who informed me that I had been very experienced in sound work in other lifetimes, and they encouraged me to assume full responsibility for my sound work at this time. It took me more than eight months to decide to do this.

I have been in a position of power four times previously in my life, and each time I made dramatic changes in order to leave it. Power for me meant egotistic "control," or abuse of the position, and it had such a bad connotation that I didn't want this kind of responsibility in my life.

However, this time the implication regarding the responsibilities of power was much greater. My sound work would impact thousands of people, so I needed to ensure that everything regarding it was done correctly, and with integrity. That was the reason why, when I realized the power of sound, I was unsure if I could do the work with the necessary humility that would be required to do only good.

One day as I pondered this issue, it occurred to me that I could act as if I was here on an assignment, and just do what my boss instructed me to do. As a result, I became comfortable talking about "guidance," because my guidance is my intuition. I know my Higher Self wants only to serve the Plan. So, as long as I don't have my little self in the way, I do well. I have developed my own unique approach, by continually asking my Higher Self the best way to do things.

Discernment About Channeling

After I became aware of channeling, I would confer with two or three channels to get their viewpoint about what I had created, even if I felt good about my findings. Even though I am not a perfectionist, when I worked with the sound, I wanted everything to be impeccable, so I still needed confirmation.

All went well until one day I found that two channels were saying opposite things regarding a small point. Their disagreement was enough to make me want to understand more about how channeling worked. I had read some of the "Seth" books years ago, and found a lot of good information there that just resonated for me.

It never previously occurred to me that the channeler can "influence" or override the information that they are receiving and giving. So, if a person has a lot of fears, they connect with non-physical beings who resonate with that vibration. And that is the type of channeling that they will provide. Much of the doom and gloom prophesies fall into this category.

The first channels I experienced had great integrity, and so I initially thought that all channels only transmitted the truth of the being they were channeling, such as Angels, Ascended Masters, Higher Beings of Light, etc. After testing some channels, I realized how far from the truth I had been. Greater discernment became the next important lesson for me, in doing this sound work with integrity.

Of course, I didn't want to reject the wonderful information coming from the beings assisting Mother Earth and helping human-kind to understand all the transformation that is happening now. I am personally open to recognizing, for instance, that the Ascended Masters are ahead of us, only because they have already mastered physicality, which is our goal here on the planet. They are very good teachers and brothers. They are also working to reach Source just as we are, and I am sure they don't want us to be dependent on them for every step we take. This would be giving them our power. We have Divine Equality as we all come from Source, and we are all returning to It, at various different stages. The only difference is the degree of awareness. Even the Christ said to us: "What I have done, you shall also do, and more." He was acknowledging us as equals, just at different stages on the path of return to Source.

The objective is to empower ourselves, because we are a part of All That Is. All beliefs or tools that we use are only stepping stones until the day we transcend the need for external help and recognize the power within ourselves.

Paying Attention to Myself

I recall one afternoon I went into my bedroom to get ready to attend a meeting, and suddenly I became completely exhausted. I was surprised, because this was not at all like me. I have always had a lot of energy and I don't remember ever needing a nap in the middle of the afternoon. Nonetheless, I was too tired to ignore my need for sleep. I lay down thinking I would get up in ten minutes, but when I tried to get up I could not walk, the pain in my back was excruciating. Startled, I went within to understand what the problem was.

The answer immediately came back that I had procrastinated for a long time about putting together a brochure containing information about my sound work. I was told that now was the time to do it. The moment I agreed to do my brochure, all the discomfort and inability to walk disappeared.

Because I was vibrating at a higher rate than before, I was much more sensitive to my higher guidance. After this experience, I determined to continually ask if whatever it was that I had planned would be serving the greatest good. I quickly learned that each time I followed my guidance, everything flowed. Life became easier and much more fun. It is like having someone that sees everything telling you what to do or where to go. It is always for the greatest good.

We all receive higher guidance, but it is our individual choice whether or not we tune in to it. So of course, anyone can have these experiences. Sound therapy certainly makes it easier to connect with our Higher Self, once we decide that's what we want to do, because the increased vibratory rate helps to raise our awareness.

The Power of Resonance

Learning About Frequencies

From the start, I knew deep within myself the power of sound. And as the sound helped me so profoundly, I became even more convinced. In addition, I remained tuned to my higher guidance, so things began to really move along.

I became more and more aware of the tremendous power of sound, both to create or destroy. The same principle of resonance is at work, regardless of outcome. For example, when a singer produces a sound that exactly matches the frequency of crystal, the crystal glass will shatter. When the frequencies match, the glass will first begin to vibrate, and will break if the sound continues long enough. Also, when a siren wails, the innate response to the sound is fear.

Every square inch on the planet is bombarded by all kinds of frequencies, such as microwaves and others. That is why it is so important to be selective, and to choose the frequencies with which we surround ourselves. It is possible to introduce frequencies which will neutralize the harmful ones, and other frequencies that go a step farther and actually enhance our well-being and repair damage. Some sounds just make us feel good. A favorite piece of music, a person's voice, or even a location, can make us feel good, because we resonate or vibrate with it.

We have the choice to create a symphony in our life, and our role is to act as the conductor to orchestrate the harmony. Every cell takes part in our body's symphony. When a harsh note is struck, like disease, we can bring it back in tune with sound. When a musician produces a sour note, we bring him back into harmony by helping him tune the instrument and re-focus attention. We do not cover up the disharmony or sour note, or remove the offender from the orchestra.

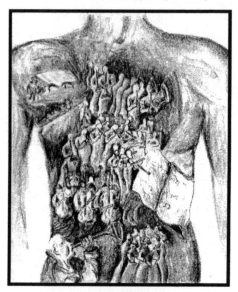

Our Internal Symphony

We often do our body a dis-service by taking drugs that cover up the problem, or having surgery to remove the "disharmony." Each cell, like each musician, is important in its divine expression and function, in order to create this symphony. In the end, it is all a question of vibratory rate, which determines whether we experience a sour note or a symphonic enchantment.

A closer look at sound will help us understand how sound can and does influence matter. I recommend watching a video titled "Cymatics," which is an interview with Peter Guy Manner. This video dramatically illustrates the effect of sound on matter. Grains of sand are placed on a drum, and when five frequencies are played simultaneously, the sand spontaneously moves to create three dimensional shapes. The sound acts as a catalyst for the creation of matter. This is a graphic illustration of "In the beginning was the Word."

The Principle of Resonance

Matter is organized by wave forms and frequencies. We all know that if we have two violins that are tuned exactly the same, and we pluck one string on one of the violins, the plucked string will produce a field of sound energy that will trigger the other violin's matching string to begin to vibrate and produce the same sound. This is called "resonance" and it happens naturally.

Resonance is a basic principle that affects everyone and everything, all the time. This same principle applies for a person in need of physical healing or mental and emotional transformation. The correct frequency reminds the body's energy field of its original blueprint and brings it into harmony. When we are in the presence of a person who is expressing joy, the energy field of her joy brings our own joy to the surface, so we resonate together. This is true of other manifestations of the same principle, in both positive and negative ways. The healer who is constantly with sick people often

The Power of Resonance

experiences disease too, unless specific actions are taken to counteract this naturally occurring event.

It is important to keep this Law of Resonance in mind. It is the operating principle of our lives, no matter which level we're dealing with—physical, emotional, mental or spiritual. The more we become aware of this, the more we assume responsibility for our own lives. We can consciously choose to resonate at higher levels. We do this by thinking and acting in a positive, uplifting way—instead of remaining unaware, and unconsciously responding to the negative thinking of mass consciousness. We can also choose to positively influence those around us with these uplifting, inspiring feelings and viewpoints, so they will also resonate to higher levels. Each of us can choose to positively resonate to an event, thought or feeling, or not. It is up to us. It is possible to respond to anger with love, and this is a good practice for each of us to do as often as possible.

Space, Time & Matter

Part of the fun of this game we agreed to play when we incarnated on Earth, is that we get to deal with space, time and matter. For example, where matter and space are concerned, these two things cannot occupy the same space simultaneously. And everything that we create has an energy field, and needs a certain space in which to exist.

Time is an even more slippery concept. Because we perceive things on this planet through our five senses, it takes "time" for each of us to process any input. The brain and the nervous system operate sequentially. For example, if we touch something that is burning hot, it may take up to a second of time before the information registers and we disconnect the contact. Everything must be done in sequence, and this takes time.

The same is true of our linear perception of time. We agreed to come into this dimension to learn different lessons. And in order to do so, a variety of frameworks had to evolve. The linear perception of time is one of them—it's a way we are able to experience this dimension.

During my near-death experience I was in another dimension

and experienced a review of all the major events of my life, going through the tunnel, and communicating with other beings of Light. All this occurred simultaneously, and took only a minute in earth time. However, back in this dimension, for me to just think about what I did during my NDE would take many hours. But that's the reality of this earthly dimension.

Over eleven million Americans have reported having near-death experiences, and these people undoubtedly know exactly what I am talking about. And, as more and more of us speak up about our experiences, it will change the face of the world. Imagine if the doctors would have been told stories of their patients' NDEs. Eventually, this would serve as a catalyst for change in hospital care, so that patients would be recognized and treated like spiritual beings having a physical challenge, rather than like pieces of meat.

The other great lesson that near-death experiences can teach us is the far-reaching impact and implication that just one single act can have on our own life, as well as the lives of others. During the Life Review, we are given the chance to view events from everyone's perspective, and to see these far-reaching implications of our simple acts. Granted, such knowledge might be very difficult and overwhelming as we live life day to day. Perhaps it's a good thing we are unable to see that in our everyday lives, just now.

Matter in Our Universe

Let us return to the subject at hand: matter. What we see as matter is simply the lowest vibrating form of energy in the Universe. The other realities have a higher vibration, and little by little we will become more aware, and gain a clearer grasp of them.

On planet Earth, the atom is king of the physical world. The atom is composed of electrons, protons and neutrons. It is the building block of all elements that exist in this dimension. It is an energy that contains *consciousness*, and can join with other atoms to form a molecule.

In initial observation, an atom appears solid. In truth, like all other forms of matter, it is really mostly empty space. What makes the atom appear solid is the extreme rapidity with which electrons whirl around the nucleus. The nucleus of the atom consists of neutrons and protons.

To get an idea of the *space* inside an atom, imagine a round table which has two tiny whirling particles of dust in the center. These tiny particles represent the *proton* and the *neutron*. These particles are surrounded by rows of more dust particles, which are also moving. There may be two, three, five or more particles of whirling dust in each of these rows. These particles in rows represent *electrons*. Remember, a proton, a neutron and electrons combine to make an atom.

One mile away, there is another table, representing another atom. It, too, has two tiny particles of dust (proton and neutron) in the center, these are surrounded by rows of particles (electrons) and all these particles are rapidly moving. The number of dust particles in each row determines which different element is formed. Atoms cluster to form molecules, and molecules form elements or matter.

Now, there is a mile of space between each table in our example, and the point of this is to clarify the fact that things that seem so solid are really just an illusion. The truth is, everything is mainly space between things. When you are in another dimension, you can see that everything is actually "full of" space.

Beyond the Five Senses

Our physical realm of existence is perceived through the five physical senses of taste, touch, smell, hearing and sight. However, we each have more than five senses in our repertoire, and these additional senses enable us to perceive finer, nonphysical energies.

Let us use the piano keyboard as an analogy. Imagine that a hearing impaired person is at a concert, watching the pianist play the piano. Our observer will be able to see the pianist's fingers going up and down the keyboard, but will hear only the lower octaves because of her impairment. She probably will not be able to appreciate why people around her are enjoying the beauty of the serenade so much. Of course, she cannot comprehend how much richer and charming sound becomes when the higher octaves are perceived.

It is the same when we limit our perception to only the five physical senses. These five senses enable us to perceive only certain "lower level octaves," playing all around us at all times. It is only when we use our extra senses that we are able to perceive the higher octaves, and thrill to a much grander symphony of living.

My NDE proved to me that there is something else beyond what we perceive with our five senses. When I left my physical body and went to another dimension where I was a ball of light, I was not subject to gravity, nor did I possess density. It was then that I realized that matter is basically the lowest or densest aspect of who I really am.

This profound experience also enabled me to realize that each and every one of us on the planet is here because we chose to be, in order to evolve and learn some specific lessons. If we could go into the spiritual dimensions easily and at will, everybody would have left by now. Granted, this three dimensional world of ours can be very beautiful, but it can also be a very difficult place for us to experience life. This is because it is a world of duality—of light and dark, high and low, good and bad, positive and negative. However, in this life we learn mostly through experiencing extremes, and it is this free-will game of duality that we choose to play, for a higher purpose.

What Are We Really?

Who are we anyway? What are we? What we perceive with our physical eyes is that we are only a physical body. But that is only the proverbial "tip of the iceberg." The greatest part of who and what we are is not visible. The fact that we are spirit, with emotional and mental bodies, as well as a physical body, cannot be determined through our physical eyes. And, if we don't want to accept this concept totally on faith, the wonders of Kirlian photography demonstrate that we have an energy field or aura around our body, with different colors and shapes. This aura represents part of the finer energies of who we are, which we are normally unable to see.

The more we become attuned to energy, especially the finer energies, the more we can feel it and work with it. The truth is that our physical body is only the densest aspect of who we are. It is the outcome or effect of our mental and emotional bodies.

Our physical body is constantly being re-created. New cells are formed all the time, and they are patterned not only from our DNA blueprint, but also by the thoughtforms we have about our body. Through the principle of resonance, the way we feel and think about our body literally creates the body that we have. Therefore, we have the power to change what we are!

Thoughts Are Things

In our Universe, there is everything we need in order to fulfill our dreams. Our only problem is that we exist in a dimension of duality, so some people think about and focus on greed, and others on lack. It is helpful to remember that we create our reality with our thoughts. Whether it's our physical body, or the material things we have or don't have, we are in charge. Once we recognize that we are more than just a physical body, it's easier to achieve what we want. And the first step of creating is to use the Higher Mind. The idea must come first, then the feelings or emotions about the idea, and then manifestation follows.

Our emotions affect our physical body, and we can be affected by others' emotional bodies as well. If we are impacted by an emotion that makes us feel uncomfortable, we can choose to release it and let it go. When we receive an emotional frequency that brings love or peace, it is almost certainly coming from our Higher Self, and we want to be certain to enjoy it. The emotional body has a larger and faster vibrational field than the physical body, so it is not a visible "body."

The mental body contains thoughts that are organized in a very specific way. We create our reality because energy follows thought. For example, consider the little boy that is told by his parents, "Cover yourself up and stay warm, otherwise you will catch a cold." Naturally, the child believes his parents and takes their idea and makes it his belief too. As a result, when he is exposed to cold temperatures without being warmly clothed, he believes that he will get a cold. His belief creates the thought, which creates the energy, and a cold is manifested. This cold then reinforces his belief, so he will be able to "prove" that colds follow chills. If he doesn't change this belief as he grows up, he may get hundreds of unnecessary colds. His belief and thought created the pattern of colds.

The same "belief programming" can apply to genetic problems. Because a parent has a certain problem, children often think they will be like their parent and inherit the same problem. For example, a woman whose mother and grandmother died of breast cancer may create breast cancer in her own body because of fearful thinking.

On the other hand, the same principle can create wonderful things in a person's life. Some people work in an infected area and never get sick, because they are not afraid and don't give it power. They spend time thinking and creating positive outcomes, instead of dwelling on negativity.

Since we manifest everything in our lives, it is a good idea to be aware of our thoughts so we can guarantee that "outcome will match thought." When we don't have what we like, it is important to change our thoughts in order to change outcome.

Our Mental Body Creates Our Reality

It is my reality that we are very powerful beings, but we have forgotten this truth. The truth is, our minds are ultimately creative, and our thoughts and words carry great creative power. The words "I AM" are the most powerful in the Universe, and the Universe will respond by providing whatever we make the call for. The Universe, however, is a neutral power source, so it cannot discern our "qualifiers" of positive or negative.

If we say "I am not poor," what the Universe understands is "I AM poor," because the *negative* or *not* is ignored by the Universe. Of course if we say "I am afraid," then that is what we will manifest too. If we think we will be hurt, the Universe will hear this as a desire, not as a fear, and will provide the circumstances that will bring hurt into our lives. If we are afraid of AIDS or cancer, our fear will draw it to us. Our thoughtforms, wrapped with emotion, act like magnets to bring the circumstance, person or event directly to us. For example, if enough people are afraid of a catastrophe like an earthquake, this consciousness will bring it into reality. So it is truly essential to think of all our "I AM" statements in positive ways, such as "I am happy," or "I am prosperous." Re-think our daily thoughts with this in mind. Becoming aware of the existence of our Mental Body is therefore very empowering.

We make choices, both consciously or unconsciously, regarding the experiences we want in our lives. And we choose experiences in order to learn the lessons we came here to master, in this school called Life. We have chances to try to learn, again and again. When we don't pass the test, we simply attract another person or circumstance to help, until we graduate and understand the needed lesson. I believe we choose our parents before we incarnate, in order to learn the lessons we most need to learn. But I also believe that at any time we have the power to resolve the problem, as long as we understand the lesson and have healed ourselves.

There is a Universal Law which says "what we resist, persists." By resisting something, we give it power. Years ago Mother Teresa was asked if she would join the anti-war protest. She replied that she would not do that, but she would add her energy to a "pro-peace" undertaking. Mother Teresa understood this great truth: the solution is to concentrate on what we want to achieve and perpetuate, because what we focus on expands.

Thus, in order to change anything, the first step is acceptance. It is empowering to recognize that we create a situation or challenge for a specific purpose. Perhaps the purpose was to develop compassion for others experiencing the same problem, or to become stronger. Every challenge contains a gift within it, without exception.

Even our so-called "enemy" is there to force us to seek the truth, to become stronger. Imagine, if you will, that your soul is gathered with other souls before your up-coming incarnation. Think of this as a "Hollywood casting call" where the director asks for volunteers to play various roles. He asks, "Who will volunteer to play the heavy-duty villain?" And a soul volunteers for the assignment.

The truth is, we have each chosen different roles before we came to earth. I am very grateful to those who volunteered to play the role of "bad guy," because it is not something I would like to do. But because we live in a world of duality, we do need individuals to play on both sides. The key is to raise our awareness so we can see that they are merely two aspects of the same thing, two sides of the same coin.

How can we recognize the Light if there is no Dark? Is the wind an enemy of the tree, or a friend? It blows the old leaves off, yet it also makes the tree pliable and strong. Little problems prepare us for the bigger ones, until eventually we see the perfection in everything.

The Power of Belief

It is important that we allow ourselves time to adjust to change. Even a positive outcome can create confusion, if we are not prepared on all levels. Studies have been conducted to see how people are doing a couple of years after winning millions in the lottery. The vast majority of these winners have lost everything and returned to their previous ways of life. This clearly proves that having millions does not create a millionaire. We have to *think* like one, and *feel* like one. In other words, be ready on all levels, so we can actually *become* one!

This, of course, applies to everything about our lives, not just having financial abundance. If we think negatively and fear disease, then we will get sick, even if we eat the most ideal food and follow a perfect diet. Remember, our thoughts carry tremendous power. Each thought has its own vibration, and this thoughtform or "cloud of vibration" hangs around us, and has tremendous impact on us. If we think negative, low-vibration, fear thoughts, we're literally planting seeds to harvest later. If you plant watermelon seeds, you won't be harvesting strawberries! The old computer expression "garbage in, garbage out" is applicable here. It is a natural law. So it is essential to monitor our thoughts and make them positive ones.

I know of people who have been healed of life-threatening disease by using specific techniques and methods, or they have been helped by gifted healers. The healing, however, is permanent *only* if their belief system and the thought-pattern that created their problem is resolved. However, if the thought or belief that created the initial problem remains unchanged, another problem will surface, and then another and another. Someone else can facilitate our healing, but we alone have the power to achieve a sustained state of health.

Of course, that ultimate power is within us. We now know that we are the causal factor or co-creator, manifesting everything in our lives. If that knowledge or responsibility is disturbing to us, remember that it takes much more energy to go against the flow, rather than with it. When we swim, what is easiest? If the word "flow" comes to mind, we're swimming in the right direction!

We are powerful beings, but if we doubt this, remember that we have already demonstrated a lot of power by creating all kinds of problems in our life! Now is the time to be good to ourselves— begin to see, accept, forgive and love ourselves, just the way we are: great, powerful, spiritual beings, who chose to incarnate in order to experience life in this dimension. This approach is definitely "going with the flow," and this will empower us.

How We Program Our Bodies

A thought of health or happiness will program our body to access all the life-affirming power within, just as a fear of disease or death will program our body for illness. As spiritual beings, we each have created our mental, emotional and physical bodies, in that order.

Obviously, the physical body is the densest of the energy fields which comprise our totality. Here, in this physical dimension, the majority of the time we forget that we are spiritual beings living a human experience. But even though energy is densest in the physical body, it still knows exactly what to do to keep itself in perfect order. We take for granted the fact that our heart continuously beats throughout our life without any conscious thought from us, yet this is a daily miracle. We also take for granted that our physical body is self-healing; if we suffer a wound, our body will heal. All of this is possible because there is an awareness in each cell, each subatomic particle, that just "knows," and so it follows the innate pattern for health—unless we override this with our fearful or negative thoughts.

Additionally, there is a field of energy that extends beyond the body itself, which we call the "emotional body." It transmits and receives all kinds of information. We are affected and impacted by whatever is around us, and in turn, we also affect and impact everyone and everything around us, either negatively or positively.

Therefore, it is helpful for us to remember to become aware of the natural beauty around us, thus programming ourselves with life energies. We can choose how deeply we allow these experiences to affect us, like the beauty of a garden, the color and fragrance of a flower, the shape of a tree, the taste of a fruit, or the song of a bird. In this way we can benefit from the great beauty, harmony and energies of nature. It is really important for us to recognize these lovely examples of nature, and just allow them to fill us with peace and love.

The Four Bodies

"Soul Choice"

Thoughts are one of the key ways that we create events in our lives. Thoughts have tremendous power. However, there are other causative factors in our lives. A principle causative factor is what I call "soul choice." Before we incarnate, we choose what lessons to learn in our upcoming lifetime. So it is conceivable that someone could contract cancer, or become bankrupt, without the *fear* of contracting a fatal disease or thoughts of poverty. Instead, these events are used as catalysts for change and transformation. So "soul choice" is a factor, in addition to our beliefs and thoughts. Yet, regardless of the cause, we are each responsible for how we create our life and our experiences, at one level or another.

The beauty of knowing this is that it is truly good advice to "go with the flow," rather than struggle against it. We can experience much more joyful lives when we are able to see ourselves as spiritual beings, deserving only good things. Even when we experience challenges, there is always a very precious gift inside that is important to lead us to the next step. This knowingness helps us to accept these challenges more gracefully.

It is important to know that we are children of God—wanting to love, to serve and to be loved. When this is our focus, we will attract the people and the circumstances that will resonate with these thoughtforms. We, as always, can choose. Choose love!

The Power of Love

The beauty of learning to love ourselves unconditionally is that once we learn to do it for ourselves, we can easily love another unconditionally, without judgement. We don't have to *like* everybody and want to be in their energy field if we don't resonate with them. But we can still *love* them.

We are each presented with a variety of opportunities to receive help from outside of ourselves, including special tools that help us return to that state of knowing and acknowledging who we really are. It's great to use a tool, just as long as we recognize that the *true power lies within us,* and we keep the goal in focus.

There is only one power in the Universe, and it is the Power of Love. The opposite of love is fear, and fear and love cannot inhabit the same space at the same time. The more we love, the less we will experience fear, and the more we allow fear into our lives, the less we will experience love.

Can you imagine a life without fear? Imagine it, because "Fear" is the challenge that we came here to master. The truth is we are all One, and there is no separation in Oneness. Separation from each other and from God is the illusion, and fear accentuates this illusion of separation. Fear drives us farther away from each other, and from God. The ultimate fear is fear of death. However, once we know that we are not only our physical body, that in fact we are a spark of the Divine and that our essence is eternal, then fear of death is no longer a reality for us.

Once we have experienced enough of this illusion of separation, we are able to reconnect with who we really are, and then life on earth becomes a playground. The more we experience life, the faster we learn the lessons we came for, and the greater our spiritual growth. Suddenly it dawns on us that the power we have is the Power of Love, and not the love of power.

Many of us are afraid of power, because we think power is something bad. This is because people who are controlling the world are mostly using the power given to them by others to further their own agendas and to gather more "fake" power. Too often such people are driven by greed and ego, and have no concept of the power of love.

The only genuine power is the Power of Love. There truly is no other power. It is important that we now become aware of this truth. It is essential that we put aside the power games we have played so well all our lives, and realize that is over. It is time to let the Light come upon us and see the truth. Sooner or later, each one of us will come to understand the reality that *we are Master, we are One, we are Love.*

When we are able to view Life from this perspective, we recognize that the only difference among us is this level of awareness.

We realize that success, money, position, intelligence, color, beliefs, heritage—in fact, *nothing* is important enough to create a separation. In the final analysis, we will each become aware of who we really are and *be* it.

Let's Grow Together

The most important work we can do relates to ourselves and our focus on growth. We incarnate here on planet Earth to learn and to grow. Often we feel guilty if we do something that is for our highest good but requires time and money. There is nothing to feel guilty about because growth is not a selfish act. We would do well to remember that each time a person raises their consciousness a little bit, the entire world benefits. So everything we do to develop our awareness is not only loving to our self, but loving to others as well. The more aware we become, the more Light, Love and Life we radiate, and the closer to the Source we get. By loving one's self this way, we are in the flow of giving and receiving.

"Sin" Transmutes Into Virtue

Sometimes it's necessary to adjust our values and reprioritize. We're not served by dwelling on the times we have succumbed to a superficial lifestyle. When we live in the past, all negative emotions surface and guilt is one of them.

The only way to become happy with our past is to use it to change the present. After all, it is why we experienced it. The moment we take the lesson of any experience of the past and use it to change the now, everything is forgiven and the "worst sin" is transmuted into a virtue.

How can we use the so-called negative actions of the past to make us a better person now? I could write another book including all the things that I did in my life which many would consider to be "bad." One of them was after my near-death experience.

I was still with my first husband and it was time to finalize the divorce. One day I discovered that my daughter had smoked marijuana at school. I tried to understand her, but I found it difficult to imagine what kind of effect this drug had on her to make her like it so much. I had never even smoked a cigarette in my life and could not understand what people found pleasurable about smoking. My husband smoked and I endured it. I had tried to light a cigarette for him a couple of times and I ended up coughing and coughing.

The next night I had a dream about learning how to smoke. So, the first thing I did when I woke up was to ask my husband for a cigarette. I lit it without any problem. At that point, I decided that at the first opportunity, I would try to smoke marijuana.

The occasion came sooner rather than later and a couple I knew offered it to me. I accepted and felt a little effect like seeing things from another perspective, but nothing really dramatic happened.

After the divorce, my daughter came home with her friends and we smoked marijuana together. Later my sons joined us. A couple of weeks later, my daughter's "friends" robbed us of all our electronic devices and art pieces while we were gone. I felt I had "invited" this unpleasant event by participating with these teenagers doing something illegal.

I was aware how serious I had been all these years, and how heavy the responsibility had been to take charge of the whole family. Now, I went to the other extreme by experiencing all I could, for the fun or for the truth of it. So, after smoking marijuana, I knew the "truth" about it and I could look at it from my own perspective, obtain my own understanding and be my own judge.

The verdict was in: I decided there was nothing really dangerous about this drug. What counts is the intention. I felt that in certain cases, it would be better than a lot of prescription drugs that people take and become addicted to. However, people that use marijuana to avoid reality are prone to go on to more powerful drugs, and that is when it can be very dangerous.

For instance, heroin addiction is very detrimental and often causes death in a short time. A person who is addicted to eating a bar of chocolate every day most likely will never develop any disease from doing this that will shorten his life. The difference is in the consequences. The problems arise when we are so attached to the drug that we need it, so we give our power away and allow it to control us.

My purpose in sharing this with you is to confirm that we all do things that can be judged "bad." Can you imagine a mother smoking marijuana with her children? At that time, it would have been enough to have had the children taken away from me. In reality, however, I used this experience to stop judging and develop greater understanding. I knew I was not a bad person for doing that and thus I realize that others who smoke may also be good people.

When we accept and take responsibility for our actions and grow through them, they become what I call "life experiences." One of the benefits is better understanding, which allows us to release judgments of ourselves and others. The only responsibility we have is to be true to ourselves. Everything we do to grow is exactly why we came here. I truly don't see anything "bad" in this.

Another lesson learned from this experience is that in order to reach balance, we often have to go to extremes. It is like a pendulum that goes from one side to the other, until it reaches a balance or center point. Wherever we are is perfect. It is when we stop the motion that we get stuck. Everything is in constant change all the time. We don't perceive it, but change is inherent to life. As long as we go with the flow, we will find the balance point which is the *now point...* the place where Power, Love, and Joy reside.

Involvement in Sound Frequencies

Early Beginnings with Sound Frequencies

I began working in the area of sound early in 1992. I used the *Korg* tuner to analyze people's voices and discover their levels of balance or imbalance. This instrument registered twelve notes and lit up when a person spoke into a microphone hooked up to it. I would log these notes onto an evaluation form, in order to recommend the mineral recordings based on the person's specific needs. It was a primitive method, but nonetheless, it was a way to begin.

In those early days, in addition to the Korg tuner, I also used a special computer program to produce sounds, which I played to clients during their personal sessions with me. After playing the mineral frequencies needed to improve their imbalances, according to my process, I would then re-test them and see instant improvements in their voices. They would then go home with the recommended mineral recordings needed for their personal balancing. I then began to use a Spectrum Analyzer computer program to analyze people's voices, and the Korg tuner became obsolete.

Correlation Between Frequency and Elements

I believe that each note correlates to one or more elements on the periodic table. For example, if a person rarely produces the note of "C" in the second octave, which is 65.40 hertz, or cycles per second, it indicates that that person is probably deficient in zinc. The element zinc has an atomic weight of 65.37. So as you can see, the note "C" and the element zinc have an almost identical number. This means a person who does not produce the sound "C" when speaking most likely will not assimilate zinc or other elements in that range. When our bodies are unable to absorb an element or mineral, it is because we do not recognize or resonate with the specific vibration of that item. Even if we eat food that contains zinc, our body will not resonate with it, and will be unable to absorb it. This is true whether the element zinc is in food or supplement form.

This was certainly true for me. When I began my research with sound, I discovered that my own voice was low in the note "E," which correlates to calcium on the periodic table. This meant that my body did not resonate with calcium; therefore, I had been unable to assimilate it from food or supplements.

This correlation between notes and elements is still considered theory and is not yet a fully established field. I am certain, however, that future research will prove this relationship between sound frequencies and the elements.

Mineral Pairs

Some of my research is based on the work of Joseph Scogna who wrote *The Promethion*. He discovered that each element on the periodic table has a natural partner. The way they work together is based on the octave to which they belong, as well as the *valence* (the capacity of an element to combine with another to form molecules) and the positive/negative value noted on the periodic table.

For example, in order to assimilate calcium, it is necessary to also use its opposite partner, selenium. Note that this is *not* the commonly known pair, which is calcium and magnesium.

Frequently, when we become aware of a deficiency, we take a supplement to remedy what is missing. However, we often don't realize that the problem is assimilation. Even if we take megadoses, we may absorb only a little. The major portion of the ingested supplements goes to other places in our bodies, and can create problems. For example, all the years I took calcium to combat osteoporosis, the vast majority of calcium was going to my joints. Instead of resolving the problem it created yet another, an arthritic condition.

It is also important to mention toxicity with regard to supplementation. Having too much of something is equally as bad as not having enough. The key is balance. Just one mineral out of balance in the body can affect all the other minerals, and create a chain reaction. It is also important to note that what works well for one person may not work for another.

All the mineral frequencies in the Sound Wave Energy (SWE) recordings have been paired with their natural partners. So, when the frequency is played, even if it is not needed, there will be no adverse effects because the minerals are balanced. We can play a recording that contains only one mineral our body needs in order to become balanced.

When we are able to speak in a harmonious way, so all the notes are present, our body benefits from receiving all the vibrations. Then, when we eat food that contains the same vibration, the body can resonate with it and assimilate the nutrients.

My First Case History - Me!

I developed osteoporosis at an early age. My best educated guess is that three pregnancies with poor diet, plus daily exposure to radiation during my six years as an x-ray technician, all contributed to my condition. Mine was such an advanced case that I lost all my teeth when I was only twenty-four years old, after the birth of my third child. I was terribly ashamed of this for many years, until the dawning of the 1990's. Once I started my spiritual journey I began to see many things differently, and changing my perception about my lack of teeth was one of them.

In order to help my bones, I took calcium, magnesium and other mineral supplements for the next twenty-five years. I spent a great deal of money for the best I could find. Nonetheless, I ended up with extremely fragile bones. My skeletal frame became extremely sensitive to my body weight. Whenever I gained even as little as three pounds, the increased pressure on my weakened spine would cause me extreme back pain.

I also developed arthritis in my hands, and they hurt almost constantly. In 1981, I began to suffer pain in my right hip, and about twice a year the pain would become so excruciating I was unable to walk. As the years passed, the pain worsened. I am certain there were calcium deposits in my joints, hands, and hips.

For several years, every time I used kinesiology (muscle testing) to ask my body if it needed calcium or whatever, I always received an affirmative answer. Of course, I never asked the appropriate question which would have been, "Will my body *assimilate* this?"

As I began working with sound, using different frequencies correlating to the minerals, my situation improved dramatically. After only five weeks of playing the mineral frequencies, all the arthritis in my hands disappeared, and I have never had pain in my hip again. In the past I had tried magnets and a variety of therapies which didn't work. But after experiencing the disappearance of my arthritis through the use of sound, I knew I was working with something very powerful.

Pain as a Gift

Pain in our body is our body's way of telling us that something is not vibrating at an ideal rate, and that energy is blocked. Pain is really a wonderful tool to prevent us from abusing our bodies. It's a sign that says, "pay attention!" It is hard to ignore, and it is an aspect of the law of cause and effect.

When we take a pill to reduce the pain, what we do is "numb" it. We don't really resolve the cause of the problem. Often more pills are needed, until the effect of the problem is too big to manage. Then the next thing we know, we have to resort to surgery to remove it.

Too often, we have been led to believe we can resolve problems with drugs, or even surgery. However, any time we approach "dis-ease" solely from the physical aspect, either by using allopathic medical tools such as drugs, radiation, surgery, or alternative medicine such as homeopathy, chiropractic, acupressure, herbs, etc., we are doing ourselves a dis-service. If we do not address the cause, then another problem will surface once the previous one has been suppressed. We may think the second problem is different from the first, if the manifestation is different, but the root cause will be the same. We must deal with the *cause* if we are to eliminate dis-ease.

How It Works

We know that everything is made up of the same basic elements, and is connected to everything else. We are All One. Our bodies appear to be very solid, yet in truth, they consist of atoms that are "full of space." We are mostly space, with everything else vibrating at a certain frequency or rate. It is all vibration, and when every tiny part is doing its job and vibrating in harmony, we are healthy.

Now imagine that a belief, a fear, an accident, lack of proper nutrition, or some other form of stress, caused us to lower our vibratory rate. Because everything is vibration and energy, this lowered vibratory rate caused part of our body to alter its vibration, then forget its role and stop functioning properly. Now that we know the problem, what is the solution?

The best way is to remind the dis-functional body parts of their ideal resonance or frequency. This is done by sending a vibration to the body part that forgot its role. The ideal frequency awakens the faulty part, reminding it to resume its work. It is similar to the way that a tuning fork works.

Beta, Alpha, Theta, Delta States

It is fairly common knowledge that our brains produce different waves, depending on which state we are in. These brain waves are counted by cycles and measured in hertz, the same as used to measure sound. Each brainwave cycle relates to a specific state. These brainwave states are given in ranges, to allow for individual differences. From a brainwave perspective, our "wide awake, alert, walking around and doing" state is called the **beta** state. Beta waves cycle at 14 and higher vibrations per second (vps).

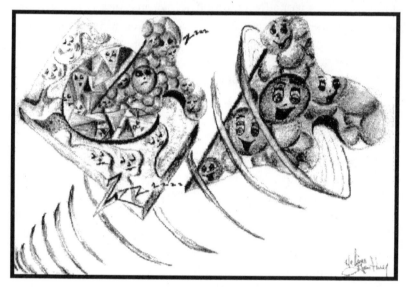

Awakening the Cells

As we relax, we are less survival or achievement oriented. We become a little "vegged out," and enter into the **alpha** state. Alpha facilitates daydreams and inspiration. Great artists, writers, musicians, etc. are all in an alpha state when their creativity occurs. Our alpha brainwaves cycle between 8-14 vps.

When we calm down even more, we enter the **theta** state. In this state, yogi masters can lie on a bed of nails, or we can walk across 2,000 degree hot coals with our bare feet. In the theta state, we're in a light sleep, deep meditation or trance, and our brainwave cycles have lowered to 3-8 vps.

The **delta** brainwave state is the slowest brain vibration recognized, and we call this state "sleep." There may be slower brain activity too subtle for our equipment to detect. Delta brainwaves cycle between 0-3 vps.

Frequencies become audible around 15+ hertz. The range of frequencies that I use in my work is 15-30 hertz. These frequencies are very safe as long as the correct frequencies are produced. This particular range will affect everything around it in a way that is manageable.

My Work in Sound Continues

For eight months I tested people and recommended the audiotapes which contained the frequencies of the different mineral pairs their bodies needed. I was using over thirty tapes, and the results were very good when clients had time to listen daily to everything they needed. Many of them needed more than twelve tapes for their various physical imbalances, yet did not know which one to use for a specific problem, such as an upset stomach. Their needs might vary daily, so those clients who were not adept at using their intuition, or using tools like the pendulum, were in a quandary on how to best use their tapes. It was a cumbersome arrangement; nevertheless, we were making good progress.

Sound Energy Research developed a very sophisticated computer program which precisely analyzed a person's voice, and which I adapted for my work. My guidance kindly enlightened me, so I began to use the program in the following way.

When a client came to me, I would ask them to think about the situation or issue they wanted to address, and then speak into the microphone for thirty seconds. The spectrum analyzer would show the voiceprint on the computer screen, in multiple octaves.

I would choose the highest point as the Power Note, and reproduce it for one minute on an audiotape, ensuring that it would be taped at the exact frequency originally produced. Next, I would find the highs and lows in the four octaves, which represented their imbalances, and these highs and lows were then played in balanced pairs.

Clients experienced the frequencies of their imbalances in the 15-30 hertz range. As these frequencies were played, causes of the imbalances were triggered and brought to the surface, so that clients could become aware and release them. Memories of physical, emotional or mental traumas were released by listening to these frequencies. Some clients ended up with more than fifty frequencies on one tape.

Sound Elicits Changes

One day I was working with a man using this technique, and I played a frequency sound related to the note of "C," corresponding to the first chakra. As I played the high and low sounds together, he experienced pain in his wrist. He said it reminded him that when he was four years old, he broke his wrist. He remembered the pain and showing his wrist to his parents. They unfortunately did not recognize the severity of the injury, and told him it was not broken and to go and play.

As a small child he interpreted this as proof that his parents didn't love him, and as a result his will to live was diminished. Playing this frequency allowed him to see a higher truth and know that, of course, his parents really did love him. Unfortunately, this truth dawned on him in his thirties, so he had carried this deep wound around inside him for many years.

I then played the frequency in the next octave that was also related to the "C" note. This time his other wrist hurt. He recalled another accident when he was a little older that reinforced the first one. He was able to remember what happened and put it in the right perspective. These memories surfaced because the first chakra is associated with survival and the desire to live. He'd had almost no desire to live, as he had felt his parents didn't love him. After he realized their love was real, the discomfort disappeared, and he was able to change his perceptions and heal himself.

More Changes as a Result of Using Sound

In another case, a woman came to see me. When I played the frequencies of the imbalances that were indicated by her voice-analysis, she began yelling and contorting, and said that unseen spirit entities were controlling her body. I was confused and didn't know what to do. I asked myself what in the world I could do for this situation. I was concerned that these entities might affect my cats, but then felt reassured that things would be fine. I played seven frequencies that released these unwanted entities. Then I continued with the other frequencies, for other imbalances. When we were finished, the woman was delighted. I had been shocked at first when the entity releases began, as it was unexpected. However, I then realized this was merely another way in which people were being cleared of their imbalances.

One of my clients at this time was nursing his mother who had Alzheimer's disease. She was bedridden, needed to wear diapers, and had no recognition of her two sons who were nursing her. After listening to her custom tape and the frequency tapes she needed, she began to get up and was able to go to the bathroom with assistance. Most exciting, though, was that she also recognized her sons! Before experiencing the tapes, she was in the final stages of the disease, and was constantly crying, yelling and shouting profanities. Afterwards, she was able to talk affectionately and lucidly with her sons. You could see dramatic differences in her behavior.

Another time, a schizophrenic man came with his father to see me. The very same evening, after the first session, this man answered his father's questions for the first time in twelve years! I became very excited about seeing the variety of possibilities for this sound work.

My only problem was that after six to eight weeks, clients needed to return to have further tests, once the effects of the frequencies could no longer be felt. Each client's visit took approximately two-and-one-half hours, every time, and I did not know how long it would take to clear all the various levels. This was a very tedious process, and it was hard for people to be in charge of their own process, as they needed to depend on me for regular testing. I was constantly searching for better ways to do my work, without the need for ongoing, regular visits. The following experience with my husband Rene paved the way for new discoveries I subsequently used in my work.

My Work Becomes Personal

The first week I had my spectrum analyzer computer program, I invited Rene to have his voice tested. Seven days later, he accepted my invitation. During the three years that we had been together, Rene had been consistent: every time I asked him if he experienced anything with any of the tools that we used, he would always reply, "I don't see any changes, this doesn't work on me."

I determined this time that I would do everything I could to ensure my husband would feel something. So, after testing his voice, instead of playing his unbalanced notes two or three minutes each as I did for clients, I played them for four to six minutes! Much to my surprise, the next morning when I asked if he'd had a good night's sleep, he answered "I feel depressed." It was the first time I had ever heard him say he was depressed. I thought that it would be a good idea for him to listen to his frequencies again. So, late that afternoon I played them all again for him.

The next day, I was in for the shock of my life. Out of the blue, he announced that I was not the right person for him. His idea of a wife was someone who would always be ready to do whatever he wanted with him. He had just retired, and so he wanted to play, walk, and swim. He did not enjoy watching me spend long hours working!

I reflected on the irony of what had transpired the previous week. I had commented to him that even though we had lost most of our money, *at least we had a nice relationship*. I realized that this was the perfect example of my reality not being his reality at all!

I remembered that Rene's heart energy center (chakra) was closed since we had been married. Of course, Rene was not unique in this regard. Many people have this chakra closed. When our feelings have been hurt, and we don't want to suffer again, there is a tendency to shut down. Men, especially, are not supposed to cry or express their feelings in this society. They keep everything inside and get terribly hurt, unable to express their feelings and emotions. Little hurts amplify and become big ones. When these hurt feelings remain suppressed for long enough, then health challenges can result.

It is no wonder that so many men have heart problems. Of course, this is how they have been raised. But sooner or later, it is important for a man to open up his heart center, become vulnerable again, and allow his feelings to be expressed. And most importantly, be able to give and receive love. In fact this is true for everyone—men and women too.

One Door Closes & Many Open

Highest Good for All Concerned

I was shocked at Rene's announcement! But as I thought about it, the frequencies that I'd played for Rene had opened his heart center. This naturally encouraged him to take action consistent with the motto "to thine own self be true." I had tested his chakras many times before and all were open but the heart. After this announcement—that I could hardly believe—I tested his heart and it was wide open.

I cried for an entire day, as I sought to adjust to the new situation. It did not take me long to realize that this move was not only in Rene's best interest, but for my best and highest good too. Therefore, I accepted the divorce easily and gracefully, because I could see what a blessing it was for me.

Rene was such a nice man, I would never have decided to divorce him. After all, I believed that I was married for good. I thought that I could be married and at the same time do the work I felt good about doing. Retirement was definitely not for me. Besides, it would have been at least fifteen years too early.

I realize now that had we stayed together, I would have been unable to do what I am now doing. Naturally, when we are in a relationship with someone, we need to take time for the relationship. From my perspective, I felt I was giving enough of my time— but apparently it was not enough for Rene. In the midst of my tears, I could see that my freedom was being restored to me, and that this freedom meant I would be able to serve more completely.

I was very grateful and appreciative of what this relationship had brought me. In addition to the wonderful internal music that I have been hearing since it began during my marriage to Rene, I also received my Green Card, which allows me legal residence in the USA. For these things and more, I am grateful.

Four days after Rene's startling announcement, I left California and drove to New Mexico to find a house to rent. Two weeks later I moved to a nice house in Placitas, New Mexico. I carted half the furniture and only my work with me. Rene kept some investments that were not completely lost. To this day we are still good friends. For me, this relationship has been perfect. It served its purpose, and by loving each other in a detached way, we were able to divorce and continue our lives without hurting each other.

Heart Opener

For some, triggering this kind of results with the recordings can be scary, but for me it was not. One door closed, but many others have opened. One day, I received a call from a lady that had purchased the full set of SWE tapes from Dr. Kupsinel, in Florida. She was very sick and was guided to work with the SWE tapes. She shared with me that when she initially received the tapes, her husband was quite unhappy with her. He thought that it was just expensive noise— the frequency sounds are not musical. However, he truly wanted his wife to get well, so after she reassured him that the tapes were benefiting her, he finally accepted them.

Unaware that the tapes had helped many animals, this same couple had planned to put their old dog to sleep, as it was very sick.

A couple of days after the wife had been playing the tapes for herself, they took their family pet to the vet. The wife called to tell me that much to her surprise, after the dog was put down her husband cried and cried. They had been married more than twenty years, and even though many close family members had passed away during that time, her husband never cried. And now it was "only a dog," yet her husband was shedding tears like a child.

She was so pleased to see her husband opening his heart. He realized that the tapes had facilitated this heart opening, and so he started to work with the tapes too. Before going to bed each evening, he would ask his wife: "Honey what tape did you play today? Which one should I play tonight?"

I have seen many couples, of every age, become much happier after using the tapes. Women are beginning to open their throat chakras and speak their truth, and men to open their hearts. People interested in using the frequencies generally want to change and eliminate their blockages. They want to release the fears, doubts and guilt, then keep their heart open to accept more love and be able to give more love.

Beginning Anew

New Mexico was a wonderful place to begin a new life. I lived in a remote area with a view of the beautiful Sandia Mountains. Such a majestic setting made it the perfect location for a retreat. I was situated on a hillside surrounded by juniper trees, high-desert flowers, and that famous big, blue, sunny New Mexico sky. This lovely setting helped me to continue my research and my work.

My two cats stayed inside, and we watched birds together through the windows. There was enough interest in my work by now to keep me busy. People were getting good results, they then told their friends, who told their friends, and word traveled. Between the work, the research, and the retreat time, I was doing quite well, just being in the flow. My main quest now was to make this powerful work simpler and more practical.

I took a workshop in Sacred Geometry in January 1993, and it answered many of my questions. I learned that sound manifests shapes that we call geometry, and that all matter consists of geometric forms. The atom creates the shape of a circle, and the Flower of Life is composed of many circles. Perhaps the circle is really the alpha and omega, or the beginning and ending, of everything. The Flower of Life and the symbol of the atom became combined to create my business logo. They are also the basis of my work.

The mineral frequencies I was using were good, but they did not address all the other aspects of a person. I knew that if the body received all the building blocks in a balanced way, harmony could be restored. But, if there is already damage because of long-lasting deficiency or toxicity, then there must be some frequencies that could repair the damage, or at least provide some help to the body.

The test I used was helping to clear the emotions stuck in different parts of the body. But, I kept asking myself, what about the spiritual and mental aspects? What if the problem is in the thought process? Are there some frequencies that can help remove blockages without having to process for years?

With all the changes that are coming very fast on the planet and in our bodies, *time* is very important. Everything is going faster compared to ten years ago. So if we are not able to raise our vibrations at least as fast as Mother Earth, we will just "check out" as many are doing. We need only observe what happens around us to realize the old way of thinking, doing, and being is crumbling away. The way things are created, they can only sustain so much tension for a period of time, then they reach a point and snap like a rubber band.

Loving Our Home - Mother Earth

Our planet is an example. We have abused her in all possible ways, with pollution of the air, water, and the earth. Add to this all the nuclear explosions that alter the earth's stability, and the depletion of Earth's resources like the rainforests, minerals etc. Is this necessary or is it greed, or thirst for power? After years of abuse, Mother Earth is responding.

For a long time, Earth could heal herself and repair the damages imposed upon her, just as we do with our bodies. But as the abuse continues, there are limits to what she can endure, the same as our bodies. The earthquakes and floods, which have become more numerous, are ways that Mother Earth is repairing and cleansing herself. And more cleansing needs to be done as the natural way of restoring herself. If she does not, we will destroy her and ourselves.

Of course, if we live in a big city, we are surrounded by manmade things like buildings, roads, cars. Life is so frenetic that we often don't take the time to look at nature. We become more and more disconnected with our Earth Mother, who is as alive as we are, and we forget what is supporting our world.

She is so beautiful and generous and loving. Because more of us are becoming aware of this and changing our attitudes toward Mother Earth, she is starting to feel good again. She is big hearted, so she is forgiving us for all the abuse, and can heal very fast as she feels our love and gratitude. Some of the big changes may not happen if enough of us continue to give her support and love. Each one of us has a responsibility to do our part, and to become aware of what is happening on our planet. Earth is here to serve us and to allow us to serve others.

Our civilization has reached a point where there are many kinds of tools available. Some are for the greatest good of all mankind, while others are for its destruction. Of course, we can fight the evil, but if we focus our energies to do that, we give it power. The other way, for our greatest good, is to use the *good* tools that are available, like global communication. Then we can network with all the people who are aware and want to make a difference in bringing about the new paradigm.

We are actually in the midst of this duality, which more resembles chaos, and these are very challenging times. But, we have the choice to be a part of this New World where we will jointly create a heaven on earth, or remain with the old way and crumble with it.

My awareness of the urgency surrounding this situation pushed me to improve my work, so that it was much easier and simpler to use. The old way of thinking, that something needs to be difficult to be good, is just that—an old belief that needs to be changed. It is a part of our old programming. If we are created in the image of the Creator, it is up to us to remove our limitations and choose what we want to create. Of course, a simple technique is good only if it works. This was my only goal. My whole life focused on finding an easier way. My ability to communicate more confidently with my Higher Self became even more important.

What's In a Name

I had been working for more than a year with the sound frequencies when I moved to New Mexico. I had not chosen any particular name for my business. I was in no hurry to identify it, because I was focusing on my research. When I began to ponder on this, the name **"Sound Wave Energy—The Way"** became very clear. It was at this same time I was debating about naming my Higher Self the "boss" and me, Nicole, as the manager or steward of my work. I really didn't want to include The Way, which is the English translation of my name, as I felt it was very pretentious to think that something can be *"The* Way."

I finally decided to include it, knowing that it would not harm anyone, but still not understanding why. It was only after I was comfortable accepting this name that I realized it was a way for my higher guidance to see if I was ready to obey and just do the work, whatever the outcome. Once I recognized this little lesson, I was released to use **"Sound Wave Energy"** by itself. I must admit that sometimes my inner guidance initially appears a little crazy.

Playing with the Numbers

Some time later I was given some information regarding the frequencies of the circle, and I realized that this was another piece of the puzzle. So, I began to work with numbers to determine which frequencies to use. I realized that each organ and system of the body has an ideal frequency. Once I identified this ideal frequency, I could play it back to the organ or system needing balance. By playing this frequency, the organ or system would remember its original blueprint, reorganize its geometric forms, and return to harmony.

I was concerned how all this would work, because I did not want to create hundreds of recordings relating to all the different body parts. I realized that I needed to address the root cause of the problem, instead of the symptoms, otherwise it would only be a band-aid solution.

Clearly, I am not a doctor, and I have no intention of playing doctor. Therefore, I do not need to know what is wrong, or the diagnosis, because I do not work with problems. My work is based on *solutions,* which naturally occur when the intention is clear and the body is balanced. By addressing the spiritual, mental, emotional, and physical aspects at the same time, this balance is achieved very smoothly.

While I was working with numbers, I discovered the frequency that would replace the test that I'd been doing—the test that showed the imbalances in people's voices. When I realized this, I laughed. I knew that, in a way, I had just put myself out of business! My sincere desire to make it simpler now made it unnecessary for me to have many private sessions with each client, and conduct the tests the way I used to. My steady clients no longer needed a test every other month. Freedom at last!

As a part of the test to identify the imbalances and the Power Note, I would ask people to speak about issues they wanted to resolve. The note in their voice that was produced most often was the Power Note, and acted as their power point to facilitate the work on the imbalances. When I found the specific frequency to replace the test, which I named the **#21** tape, the Power Note became obsolete.

In conjunction with this new system, using the #21 tape, I began finding people's **SoulNote**. This was an improvement. It didn't address problems, instead the idea was to reach for the *essence of their being,* the *SoulNote!*

What is the SoulNote?

Soul is part of ALL THAT IS. Each soul can be compared to the drops of water that create an ocean. The exception to the analogy is that when each drop is separated from the Source, it is unique, and has a specific frequency. That unique, special frequency remains the same until the drop, or the soul, returns to the Source and becomes ONE again.

When we learn our own special SoulNote, we can use that knowledge in helpful ways. For example, when we hum our SoulNote, it enables us to reconnect with Source—a good time to make major decisions. That particular frequency makes our body vibrate, and we literally become an instrument. Humming our SoulNote brings us joy because *Soul knows only joy*.

We can also use it any time we feel upset. It is easy to forget who we really are and identify with problems. When we do so, joy is no longer present and we may feel sad. Humming our SoulNote will help us to reconnect and remember who we are and just *be* it.

Many people immediately experience this joy upon humming their SoulNote. Often those who hum their SoulNote say they experience a state of bliss while toning—it is a very powerful note.

The SoulNote can be used as a tool to empower one's self when working in other areas. I recommend that people who know their SoulNote hum it just prior to playing any of the SWE recordings, after they have stated their intention. Ten seconds of humming is sufficient to bring that Soul energy to the area we work on.

We can hum our SoulNote any time. When we say mantras like "OM," "HU," or any we feel comfortable with in our meditation, saying them in our SoulNote makes our contemplative time even more powerful. Because humming our SoulNote is acknowledging our essence, it is very empowering and essential to do at this time. Even if other people are present, and they take your tone to hum, it is like saying to you "I love you." The ideal is when each person has his SoulNote and can choose to hum another SoulNote but come back to their own.

I was invited to go to Hawaii by a very gifted lady doctor named Linda Fickes who had experienced the SWE tapes and just loved them. One of her clients came to see me one morning to find her SoulNote and had an appointment with Linda that afternoon. As the client began humming her SoulNote, Linda could feel the energy field of her client expanding so much that she wanted her SoulNote also. Linda's work is so valued because she sees auras and can determine precisely what the client needs.

I have worked with each member of some families to find their SoulNotes. They were humming the respective SoulNote of each person, all together. What a wonderful way to help acknowledge and reconnect with God within each one.

Group SoulNote

One of the next challenges that we need to master is cooperation. If we are to accomplish our goals, we must learn how to work in groups, as the time when we could do everything alone is past. We need the expertise of many others, because what needs to be accomplished is so big that we cannot do it alone.

For this reason, I was guided to create an enhanced way to use the SoulNote that will help develop the ability to work in groups. When a group decides to cooperate in order to achieve certain goals, it is important to begin with a specific plan. Then, each group member can have their individual SoulNote done. The next step is to produce a tape that consists of the average frequency of all the group members. For instance, if one member has a SoulNote of 196.967 hertz, another has 107.870 hertz, and a third has 131.30 hertz, I will add these three numbers, divide the total by three, and come up with a **Group SoulNote** number, which in this case would be 145.379 hertz.

When the participants of a group meet, each member starts humming their own SoulNote and the group joins them. They then hum their Group SoulNote together. Group focus is set, unity is created, and great results can be achieved. Then the project, or the mission, becomes the goal, and each individual becomes part of a *group consciousness*. This way it is easier for the "little self" to stay out of the way.

I also produce SoulNote recordings for couples. This is a wonderful and powerful tool to use, but I only recommend it when the couple is clearly in a relationship that is already a strong one, and they are not dependent upon each other in any way, but are each autonomous individuals. Otherwise they can get "lost" in each other, and not be in their place of personal empowerment.

We seldom experience moments of complete silence, so we get used to all kinds of sound that affect our bodies continually. When we tone our SoulNote, or other healing sounds, we help to bring ourselves back into balance, since the sound comes from within.

For the majority of people, it is easy to hum their SoulNote. Others take a little time to clear some issues before they are able to resonate with their SoulNote.

For example, a past life issue could block the SoulNote experience. If a person was hanged in a past life for speaking her truth, and that situation has not been completely resolved in this lifetime, blockage might exist in the throat area, preventing her from being able to hum her SoulNote. I then play the same SoulNote for the client, but one octave lower. This frequency then triggers memory of the event stored in the cells, and the stored memory blockage is released. This is a very simple method to get rid of core issues that we carry from lifetime to lifetime.

I tested and worked with people's SoulNotes for over a year before I began emphasizing the lower octave. I realized that this lower octave brings to awareness other big core issues located in any part of the body; issues that have not been resolved from past lives. When the lower octave of the SoulNote is played, a tingling pressure, or slight pain sensation, may be experienced. No-one really likes to recall those negative circumstances, but we do want to resolve things and remove blocks in order to grow. So, as most of us want to resolve these issues, and time is limited, the faster we can release all these unseen issues the better.

The lower octave of the SoulNote is the only frequency that I recommend be listened to in a quiet place, and to focus on becoming aware of sensations in the body.

We really don't have to process these issues in order to heal them, it can be done in four simple steps: 1. Acknowledge the issue. 2. Forgive yourself and any others that were involved in the event. 3. Give thanks for the lessons learned. 4. Release the situation into the Light. That's it. If the sensation is not gone within a minute, we can ask the help from our High Self, angels, or guidance.

To achieve maximum growth, we want to clear these soul issues and then play our SoulNote as much as possible. The lower octave of the SoulNote will often resonate with the lower chakras because it is where the survival aspect of ourself is located. We may discontinue listening to this note when we have cleared all these past issues, and just play the normal SoulNote and be joyful.

Higher Octave Of The SoulNote

At the beginning of 1995, my guidance gave me the green light to put onto audio tapes the frequencies which had been given to me two years prior and which, up until now, I had been unable to produce. Our vibratory rate was not high enough to sustain these frequencies at the time they had been originally given to me. I was also given the permission to offer people the higher octave of their SoulNote.

This higher SoulNote resonates with the higher energy centers of the body and helps the connection with those centers above the head. It is, for the majority of people, an instant connection with the higher part of Self. Some people just feel that it is "heaven." The sound resembles a higher pitch that some people have already heard. As we raise our vibratory rate more, we will feel more comfortable with the sound of the higher part of Self.

When I tested a person, after I found the SoulNote I would play the lower octave, the higher octave, the two combined, and then have the person choose what they wanted. I always play the SoulNote on side A of the tape. On side B, I record the lower and the higher octaves.

For instance, if a person has already cleared the major past-life issues, the lower octave may not be necessary. Or if someone has a major blockage in the higher chakras, the higher tone may not be comfortable yet and it is important to acknowledge this. It is why the people will first experience the different possibilities before I record the sounds on their custom tape.

How The SoulNote is Found

I gently guide a person to invite their Higher Self or their I Am Presence to come upon them so that they will produce their SoulNote, which is the frequency that is the most prevalent when a person speaks from the heart. Then I will invite them to speak into a microphone for about one minute. A spectrum analyzer will reveal the SoulNote. They may speak on any subject that expresses joy or sadness; it doesn't really matter, because if it really comes from the heart the SoulNote frequency will be there.

It is important that it doesn't come only from the logical mind, because what we will find will be the brain note and we have plenty of that one. I found that it is preferable for a person to speak in their native language, or at least the language they feel the most comfortable with. I don't need to know what is said, as long as it comes from the heart.

Since our SoulNote is produced only when we speak from the heart, it is very important that we try to speak from the heart as much as possible.

Because the testing is best done in person, I will soon be training others to find the SoulNote.

What Is a Chakra?

One morning, during my reflection time, I received guidance that I should work with the energy centers of the body called *chakras* (a Sanskrit word meaning wheel). At the time, I knew very little about chakras. I learned that each person's body has seven major energy centers—whirling vortexes of energy. These chakras are powerful energy fields which are invisible to the eye, yet nonetheless are quite real. Each center is connected with one of the seven ductless glands in the endocrine system, and each chakra's role is to stimulate the associated gland's hormonal output. These hormones regulate all the functions of the body.

The chakras, or energy centers, are the conduit of life force from our I AM Source, or Divine Energy. This life-force energy moves at a very high speed and must reduce its vibrational rate in order to merge with the cells of the body. These electrical impulses are transformed into chemicals upon entering the different glands. This Divine Energy first enters the chakras, then goes through the energy channels of the nervous system and the endocrine system. The endocrine system uses the various hormones to send energy to the bloodstream and to all the cells in the body.

It is tremendously important to keep the chakras, meridians, and nervous system in balance, in order to facilitate the free-flow of divine energy. In a healthy body, each of these vortexes whirls at great speed, thus permitting the vital life energy, called *prana*, to flow upward through the endocrine system and maintain the body in a perfect state of balance. But if one or more of these vortexes begins to slow down, or get out of balance, then the flow of vital energy is inhibited or blocked. This causes optimal health to decline, which then manifests as dis-harmony or dis-ease. Thus, aging and physical deterioration set in.

The Business of the Chakras

The chakras are to the physical body what different departments are to an organization. Every organization is created for a certain purpose, and this purpose can be stated in a Mission Statement which is meant to sustain the original intention or focus throughout all business dealings. This Mission Statement intention in ourselves corresponds to the 7th, or Crown, Chakra.

Chakra #6 - the Third Eye Chakra, corresponds to the administration department within every organization, where projects are accomplished and a set of rules follows.

Chakra #5 - the Throat Chakra, corresponds to the communications or public relations department—the outreach department.

Chakra #4 - the Heart Chakra, represents the manner in which the service of the organization is performed. This relates to how the employees feel accepted, loved and appreciated, and how the services provided make customers feel good and want to return.

Chakra #3 - the Solar Plexus Chakra, corresponds to the inflow, and the fair exchange of money for goods or services to the customer.

Chakra #2 - the Sexual/Creative Chakra, represents the development or creative endeavors department—creation of the service or product, and the energy needed to make it.

Chakra #1 - the Root Chakra, relates to the department within an organization that serves as the foundation, and creates the stability, order and maintenance, necessary for success.

Obviously, each department is critically important and needs to be aligned with all the others. If only one is not doing its job, energy will not flow freely, and failure will result sooner or later. No single department (or chakra) is more important than another. When an emergency arises in one department, if all are aligned with the ultimate mission, all will respond rapidly to the emergency, and move in concert in the same direction. It is not the

Chakra Balance

ideal situation when one department says one thing, and another something else, and off they go in different directions. Then things become more difficult, for either the business organization or the human body. Good clear communication between the groups is essential for health or a successful business.

How I Put My Studies to Use

I read books, researched and continued to study about the glands in the body and their needs. Yet at the same time I was very aware that I was receiving new information that was not in the books. It was an interesting way of using both left and right sides of the brain simultaneously. The chakras were my answer. I realized that this was how I could use the frequencies in a manageable way, without too many recordings.

So it became my challenge to put the correct frequencies together to include everything needed for each chakra. This was a marvelous study of the physical, emotional, mental and spiritual bodies, to put all of these frequencies together in such a wonderful way.

When I began to mix the different frequencies that correlated to various minerals, noble gases, vitamins, amino acids and hormones, for each chakra, I was naturally very curious to see how powerful they would be, and how well they would work.

I worked intuitively for many months, and I finally had a complete series of tapes that I was ready to test. I *knew* the frequencies would work because I had total trust in my guidance. They were to become the foundation or building blocks for my work. My new **Energy Center (Chakra) Series** of tapes was a very important step in helping people return to harmony. It was clear to me that I was only a steward here on earth, to bring this sound wave information to others who could really benefit from its power.

Miracles Begin

Giving the Frequencies a Try

I began to play the **Chakra** recordings and other new frequencies I had developed. For the prior eighteen months lots of people had worked with the frequencies that only correlated to the minerals. Now it was time to try this dramatically expanded group of frequencies.

I was in pretty good shape at that time and so I didn't personally detect any noticeable changes, but the first reaction came from one of my cats. Remarkably, Tiger's strong male energy became softer; he no longer protected himself so much, and adopted a more open, friendly, and loving demeanor. The change was pronounced.

Changes for Robert

Next came my son Robert! The main reason I had spent decades searching out alternative medical approaches was so I could find something that would help Robert. As mentioned in Chapter One, his entire endocrine system had shut down when he was five years old, no doubt a result of the radiation exposure we both suffered during my pregnancy with him.

Over the years, Robert had tried many different things that I recommended to him. Some had helped a little, but nothing really dramatic transpired. For years he had been taking extracts of different hormones and this had helped him grow to some extent and reach a certain height. But without the growth hormone to help him mature and develop normally, he looked much younger than his true age. Even now, in his thirties, he is still asked for proof that he is over 21. When he was younger he was unable to accept this situation very well, and became quite angry and disgusted with anyone who would not permit him to enter their establishment unless he showed his ID.

Robert's problems were hormonal. When his glands didn't produce their specific hormones, his body's systems were left without direction. It would be like running a company without anyone in charge to direct the different departments. Without direction, deterioration is rapid. It was essential for Robert to take hormones in order to remain alive; yet once, he deliberately stopped taking them.

Gradually, Robert's life had become so lonely that he wanted it to be over. So he stopped taking his hormones and became very, very sick. It took eighteen months of tender, loving care to bring him fully back into this world. During his recuperation, Robert discovered that he had a great writing ability. So after his recovery, he published his first book. Even in the bleakest of circumstances, blessings can be found.

Since I became aware of alternative approaches, I began searching other areas than medication to correct Robert's condition. I tried the Rife frequency generator, crystals and magnets, and I became a hands-on healer Reiki Master. These different modalities were helpful, but only marginally in Robert's case.

A Visit from Robert

When I began to work with frequencies, I was so amazed at the results people were having that each time I spoke with Robert on the phone, I told him of the testimonials I was receiving from people who were using the tapes. However, each time Robert's answer was the same, "I am not interested. I have tried many things, and it is not worth it."

Then one day in April 1993, after I had completed all the research work to create the **Chakra Series** of recordings, I received a phone call from Robert saying that he would like to visit me. We had not seen each other for more than a year, so I was very excited about his coming. Naturally I was not only anxious to see him, but also to convince him to try the frequencies.

During my very next meditation, I asked what I could do for Robert during this visit, and the answer I received from my Higher Self was: "Don't expect anything." My reaction was "What? I know that I have what he needs, yet I cannot expect anything?" At first I could not understand. Of course I had expectations! How could I achieve the state of having none? I was baffled.

Unconditional Love

I spent many hours trying to figure out how in the world I could comply with the directive "don't expect anything." After a month of thought, I realized that the only way I could enjoy Robert's visit, and not expect anything, was to treat him as if it was his last day on earth. How would we act knowing that a person we loved was spending the last day with us. Naturally, there would be no expectations, nothing to change, nothing to do, we'd just relish the time and love unconditionally. So I tried it.

I remembered how Robert liked to play when he was little, and how as a busy working mother I was almost always too busy to play. I determined that the two weeks we'd spend together would be dedicated to whatever Robert wanted to do, without any expectations on my part. I went to the store and purchased games—we never played them, but I was ready! I also promised myself that I would make no appointments during his visit, and do my frequency research only when Robert was occupied doing other things.

When I picked him up at the Albuquerque airport, the first

thing he said was that he wanted to see my computer. Since he is a writer, his computer had become a valued friend, and he wanted to meet mine. So when we arrived home I showed him my computer, and he immediately asked me to do his SoulNote test. I truly believe that this occurred because I had released all attachment to outcome, and had surrendered.

I analyzed his voice to find his SoulNote and his imbalances, and then played those frequencies for him. He did not want to hear his SoulNote, so after he'd listened to all the notes related to his imbalances, I asked him if I could play the Chakra recordings. He replied "Sure, play what you want."

I was so surprised, that without even considering if it was appropriate to listen to more than one Chakra tape at a time, I began to play two tapes at once on my two subwoofer systems, which produce the lower frequencies and give faster results. Very shortly I noticed that Robert began to open up. Robert considered himself to be a loveless person—he felt hatred toward himself and others. It took two days of intense discussion to help him see and accept all the love that he had inside himself. His love of animals and all nature were clear examples that he could see for himself. We also focused on the many challenges that he had faced, and agreed that not too many people would be able to respond with love when they had been put down by others all their life.

What transpired was a miracle. Robert began reading different books to help understand himself. He began applying techniques like hug therapy when we visited people. He meditated and truly became connected with who he really is. It was a wonderful blessing for me to observe all this. If I had expected anything, I would never have expected all this. It was like watching a bud that had resisted opening for a long time, and then suddenly when you no longer expected it, burst forth into full bloom. The beauty of such an event is so awesome that words can hardly describe the magnificence.

Love is the universal solvent. When the heart is truly open, *everything* is possible. This fourth chakra, the energy center of the heart, is the center of our bodies—and eventually it will be in the center of our lives. It is the key. We each have this jewel within us, and we each have the choice whether to use it a little or a lot.

When Robert returned to Montreal two weeks later, his whole attitude had changed. He was able to completely stop taking the hormone extract we thought was essential for him to live. He decided to volunteer at the Children's Hospital where he'd spent so much time when he was little. More importantly, he began to express feelings of love to himself and others. He placed positive affirmation cards on the walls of his apartment to help him sustain his new happy feelings. He became happy and healthy, and developed his inner senses so he now communicates with his own Higher Self.

The next holiday season we all gathered in Montreal, and had a wonderful time—better than ever before! Robert demonstrated a beautiful truth to each of us, that when we love ourselves, we can easily accept love from others. What a wonderful gift.

Synchronizing Both Hemispheres of the Brain

Prior to Robert's visit with me in New Mexico in May of 1993, I was still organizing all the specific frequencies I had received to enhance the other bodies. These were to become my **Physical**, **Mental**, **Emotional** and **Spiritual Series**. The new Chakra Series was one way, but I was still doing extensive testing for imbalances, and this was a cumbersome and lengthy approach.

In April 1993 I attended another seminar on Sacred Geometry. By now, I had learned to follow my guidance, so before departing on my trip I checked again to see if I was doing the right thing, and I received a big "Yes!" So I flew down to Dallas, Texas, and was delighted with all the fascinating information that Drunvalo Melchizedek provided during his five-day Sacred Geometry workshop.

On the fourth day, I received confirmation about the music that I had been hearing in my head. Drunvalo told us that many years ago, the Rosicrucians conducted a study. They had literally weighed human spinal vertebrae, and discovered that the ratio of weight between each vertebrae was the same as the ratio between the notes of the chromatic scale. This difference in weight corresponds to the different frequencies of the harmonics of music. When both hemispheres of the human brain are synchronized, in harmony and balance, they create a vibration in the spine, which is heard as music inside the head—often known as the Music of the Spheres.

*Acoustic Brain Research** produces a music tape titled "Wave Form" that is based on this theory, the principle of the Harmonic Fifth. The music is tuned to the human spine, and helps to integrate the right and left sides of the brain. This is the *same* music that I have heard since 1990. The only discrepancy is one note which is slightly different, and is probably due to the fusion of two of my spinal vertebrae. After listening to this tape, I realized that this was not only "my music," but could be heard by anyone that synchronized both hemispheres of the brain to work together in harmony.

When I pondered why I had not understood that this Music of the Spheres was for everyone, I received the thought-impression, "You cannot receive an answer if you don't ask the question." Our inner guidance cannot interfere with our free will, and cannot give us something if we don't request it. My behavior pattern of *not* asking, which I had learned in childhood, was now obsolete. I had to develop the habit of *asking specifically* for every special thing that I wanted.

I feel so blessed with all my needs being fulfilled. I have music in my head, good health, work that I love, and hundreds of wonderful people that I continually meet, who then become friends. Sometimes it is hard to request more, even if it is not for myself.

Becoming Mobile

I was used to receiving information in my quiet times. Some of the most meaningful information I received was that I would be putting together a center where people would come to achieve a higher spiritual state. This center would use sound, light, aroma, color, crystals, and sacred shapes, among other things. The whole idea was very exciting.

---

* Copyright 1989 Tom Kenyon; Acoustic Brain Research, Inc.,100 Europa Drive, Suite 430, Chapel Hill, NC 27514

The lease on my house in New Mexico would expire at the end of June, so as soon as Robert departed for Montreal, I began to search for a bigger house where I could start the center. Every place I looked at was unsuitable. One afternoon, after viewing yet another unsuitable house, I felt an impulse to overtake the trailer that was traveling in front of me. I got an immediate hunch to "Look at that." I was startled! I began this little conversation with myself because previously I'd never even wanted to pull a little trailer behind my car. Now, I was debating with myself about pulling a little house behind a truck! All in a moment's time! Being a Sagittarian, I have always liked adventure, and moving has always been fun for me; but being on the road by myself would be quite another thing.

It later occurred to me to investigate buying a trailer in order to take this specific sound technique to people in many different places. What a great idea! So, after consulting with my ex-husband, Rene, who had first-hand experience with such things, I enthusiastically decided to "go for it!" I actually began to shop for a motor home.

Trust and Total Faith

When I applied to the bank to borrow money to buy a used motor home, I was turned down. So I came back home and had another conversation with my Higher Self that went like this: "I cannot get one. I don't have the money." And the answer was: "You have credit cards." I responded to my Inner Self: "Yes, but I've never bought a big item like this with a credit card!" Then a little later, "I will do it, but I need help to pay the bills!" Two days later I found a ten-year-old motor home, for which I paid "cash." The couple from whom I bought it drove home with me, and I asked the husband to help me hook my car behind the motor home, using the easy-hitch I'd had installed on the car.

The very next morning I was ready to leave for Dallas, Texas. I got everything ready in the motor home, including my two cats, who hid under the sofa. I turned the ignition key, put it in gear, and began going down the hill. Bang! was the next sound I heard. I immediately opened the door to survey the damage and identify the problem. Apparently the car had not been properly connected, and had banged into the motor home. Fortunately no real damage had been done to either the car or the motor home, since I had been driving so slowly. When I opened my door to get out, the first one out was my cat Tiger. He simply flew outside as he was so afraid of the noise. I called him, but he did not appear—he was gone.

As I looked down on the driveway, I noticed a yellowish kind of stick, about three to four feet long. I did not remember dropping anything like this, but suddenly I realized it was a snake, moving in the same direction as Tiger. It was my first encounter with a snake, so I put on some boots, and went in search of Tiger. I thought he might have gone to the little creek, but I could not find him. So I returned to the car, hoping he would come back soon. After reading the instructions about the hitch, I successfully fixed it. By now it was ten o'clock in the morning.

I had planned to be in Dallas the following day, but I didn't know what had happened to Tiger. He hadn't returned and I did not want to leave him there, so I decided to wait in the motor home, ready to go. Eventually Tiger showed up, late in the afternoon. I was greatly relieved, but felt it was too late in the day to depart, so I did my very first camping in the driveway of my home! My house was completely empty, so I had no choice but to christen my motor home there. What a way to start an adventure—I wanted to leave, but I couldn't. I'm sure there was a reason, but no-one has shared it with me yet.

A New Style of Traveling

At five a.m. the next morning we were on the road. I drove the speed limit in the slow lane on all the freeways. I carefully assessed where I needed to stop, because I could not back up with the special tow hitch. Since I was alone, I had to check everything—the map, the car behind, the cats, and anything else that was pertinent. I made certain I put the cats in the bedroom before leaving the motor home, so they would not run outside again.

Since I had left so early in the morning, I reached my planned RV park too early to stop for the day. So I continued on the road. Once I was going, I liked it. The motor home was easy to drive, and

I was seated up high and could enjoy the panorama more than if I'd driven my car. The main thing that I needed to be aware of was the height and length of the motor home plus my towed car.

Once, I missed a turn and thought it would be safe to turn into a school entrance in order to go back the other way. But once in, I realized workmen were paving it, and the other end was closed. Since I could not back up I had to unhook the car, turn the motor home and then the car around, rehook everything and get going again. It was not much fun, but I learned quickly.

At fifty-five miles an hour it took me fourteen hours to get within forty miles of Dallas. I decided to stop, as the headlight beams were not set very high and I couldn't see very far. The chivalry of the professional truck drivers was truly appreciated, as I learned the challenges of driving with such a long vehicle. They were polite and considerate, and would give me clear signals to pass, or return to the right hand lanes, as I traveled along with them.

Night came fast, and having found nothing else, I stopped at a rest area and settled near the big trucks for the night. The drivers had been so considerate throughout my first day that I felt safe parking between their trucks for the night. My cats were happy to finally resume their "normal" life. The entire time we traveled together, the cats hid until I stopped driving; otherwise they were very good companions.

Another Tiger's Story

Once I arrived in Dallas, my friend Barbara invited me to park in her driveway. It was very pleasant being able to stay there and test people to find their SoulNote. My only problem was I could not run the motor home's air conditioner, since it was 220v current and was not compatible with the 110v current that Barbara's home had to offer. My computer couldn't make it in the Texas heat, so Barbara invited me to test people inside her home. I moved all my equipment into the den and worked from there, still using my motor home for my cats and myself to sleep and to eat.

Barbara had arranged lectures for me, and brought people to me to work with. I was very busy. The first morning after moving my equipment into her den, I vacated my chair for a moment and Barbara's cat, also named Tiger, took possession. When I saw this, I pulled up another chair for myself and left him alone.

I played the frequencies for twelve hours the first day. The only time Tiger departed from his chair was when I played the tape **Brain, Courage & Prosperity.** I decided the cat didn't need to change his beliefs and was happy with his prosperity.

Now, Tiger was a nineteen-year-old cat, which is considered very old. His skin was full of scabs and he was very skinny despite the fact that he ate three 6-oz cans of cat food daily. He was unable to get to the litter box, because he had such difficulty walking.

After five days of hearing SWE tapes, all of Tiger's scabs had disappeared, he had reduced his food intake to one can a day, and he could get to the litter box in time because he could now walk easily. What a difference! Animals instinctively know what is good for them, they have no resistance or beliefs that will prevent their healing. Animals don't think "This can't work." But sometimes we as humans certainly do. This is why it is so important that we address the mental, emotional and spiritual aspects when we have a physical challenge. The body is designed to repair and heal itself, if we allow it.

Tiger's rejuvenation was convincing proof that these tapes help because they emit healing frequencies, not because we "think so." It eliminated the placebo effect hypothesis. They helped because they resonate with the different body parts that are perhaps "sleeping" and not working properly.

One of the residents of Barbara's home was a renowned doctor. He saw, first-hand, the results of Tiger's rejuvenation, and also the many wonderful changes different people experienced. As a result, he invited me to be the guest speaker at a dinner meeting. This was a group of people who gathered monthly to share new information and discoveries relating to alternative scientific issues. I was a little reluctant to go, because I felt that I didn't have a lot of experiences to share, but I had a strong hunch that I needed to be there, so off I went.

Dinner Talk

Dinner was scheduled to begin at 6:30 p.m. and end three hours later. By nine o'clock several different people had spoken, yet I hadn't said one word. Then my guidance gave me a nudge, it was time to speak up. The current topic was about health food stores that would no longer be allowed to sell highly beneficial products. Very casually I said, "We don't have to worry about this because we can use the *frequencies* of these supplements, and our bodies will assimilate them all from the foods we eat."

This was sufficient to get everyone's attention, and when I spoke about the correlations between the frequencies and the atomic weights of the elements, the facilitator for the evening began to laugh. He laughed for so long that I stopped talking, not knowing what to expect. I prepared myself for any outcome—ridicule, even a tomato or a rotten egg, amusement, or a big *aha!*

It was the latter, because he had pondered the same possible correlation himself. I was delighted with his response, and so I shared the experience my other son, Francois, had suffered the previous year. I told the group that in 1992 Francois had become extremely sick, ending up in the hospital. It was discovered that his spleen was eating the new blood cells his body was producing, and his hemo-globin was very low. He endured numerous tests, but the doctors could find no cause of the problem, so they resorted to prescribing cortisone to arrest the situation. He was put on 25 milligrams of cortisone per day, and every time the dosage was reduced, the problem flared up again. Francois had been on this medication for nine months when his wife Nicole came to visit us in Vista, California.

Francois had recorded his voice on tape, and Nicole brought this tape with her. I analyzed his voice and produced a tape for him to play nightly. It was impossible for him to play tapes during the day, because at work he was constantly on the move, with no chance to plug in a tape recorder. So Francois purchased a tape deck with auto-reverse, in order to listen to the frequencies throughout the night. At that time, I had only developed the frequency tapes which corresponded to the various mineral pairs, but I also included the cortisone frequency on his tape, in addition to the minerals that needed to become balanced.

The following month he went for his regular checkup at the hospital, and the doctors worked again to reduce his cortisone dosage. This time it was successful. He was able to reduce his cortisone intake from 25 milligrams to 15 mgs for a month, then down to 10 mgs the next month, down to 5 mgs the following month, and finally to 2.5 mgs and then none.

The man in charge of the dinner group was quite impressed because he had a similar problem to Francois'. After I completed this story I was encouraged to tell other stories, and then I spoke about the SoulNote. The group was so responsive that we remained at the restaurant until 10:30 p.m., and then ten of the group wanted to come back to the house to see my special computer program. They stayed until 2:30 in the morning! It was so exciting to see such sincere interest and the enthusiasm in their eyes and smiles.

Expectations Adjusted

I had attended the dinner meeting with some trepidation. My resistance had been based on a few experiences I'd had sharing this technique with very left-brained people. I simply could not satisfy their demands for scientific data with details of the why and how it works; they could never have enough information. The quest for information can be a drug that prevents one from getting in touch with feelings. These prior experiences had made me reluctant to go to this dinner, but I was committed to my work, so when I received the clear command from my guidance, I obeyed. I realized that night how much I really felt I was a steward of this work, because even if I had been laughed at during the dinner, I would no longer have taken it personally.

One thing I now realize is the importance of developing ways to bridge the scientific and the spiritual communities. Many people are joining together, with the common goal of merging science and spirit. The separation we used to think was so real is proving to be only an illusion—an illusion that will soon be history.

In Austin

The next place I planned to go was Austin, Texas. I had a chance to expand my own love when I arrived there.

I was not yet familiar with RV campgrounds. I didn't know that some people lived there permanently, for years. So, the first time that I opened the door of my motor home after parking there, a small resident cat, half the size of mine, was outside watching what was going on. My cats and I were apparently on her territory. I didn't know that female cats are as territorial as males. To my great surprise, when my Tiger came out, she ran after him—and he disappeared. Tiger had been a fighter. There were no cats too big for him. But since I'd been playing the **Chakra #4** and the **Love** tapes, I had noticed a change in his attitude. He was no longer inclined to fight and preferred being more loving.

When Tiger didn't return after a couple of hours, I searched the campground but couldn't find him. At the end of the day, no Tiger. I sat and started to do my meditation, not knowing where he was. In the middle of my tears, I decided to let him know that whatever he chose would be okay, and that I released him. I thought that maybe somebody liked him so much that they had decided to keep him. I thought about him a lot during the following days.

To be sure I would be there if he returned, I decided to redecorate my motor home. I changed the blinds, I made beautiful purple drapes and changed the wallpaper. The color brown was predominant in my motor home and I wanted to change it to lighter and more vibrant colors.

After eight days of waiting for Tiger, I received guidance that he was not dead but starving somewhere in the RV park. This was enough to make me cry again and to start an active search of the campground. I showed his picture to people, searched around and under buildings and spent two nights watching out for him. Two days later I realized what the message behind the situation was for me.

Before Tiger and Princess, I'd never had a cat that I was solely responsible for. I had a tendency to see them as animals that needed my care, but I'd never paid much attention to what they were giving to me. I loved them but I couldn't fully understand the relationship that people had with their pets. At that moment of recognition, I truly embraced Tiger and all animals as very important beings.

Later that same night, I became aware that a cat was walking on me as I slept. I had a little door under the bed that opened to the outside, so the cats could come and go. When I turned on the light, I saw that Tiger was back. He still had the same powerful, calm and loving manner, he was just skinnier. His adventure didn't change him—but it had changed *me*. I thanked him for being so patient, starving for ten days in order to give me a chance to experience compassion and humility. I realized that animals have a lot of wisdom. They always remain in touch with who and what they truly are. They are here to help us remember the same things. They love us just the way we are, all the time, without wanting us to change. This was Tiger's lesson of unconditional love for me.

On the Road Again

The Lecture Trail

One of the greatest blessings of doing Sound Wave Energy work is that I get to meet the most wonderful people. It doesn't matter if I'm talking to a group or doing private SoulNote sessions, I never feel like it's "work." The only part of this whole thing that feels like work is when I have to carry my computer with me when I fly. I have to carry a desktop computer—not a laptop—because of the nature of the program I need to use. So I drag my computer plus a small television monitor, and believe me it's quite heavy, so *that* feels like work! But that's the extent of it!

I just love sharing SWE with everyone I can, and although I often have a full schedule and begin my day at 8 a.m. and keep going all the way to 10 p.m., I still feel great when I am done. The joy and satisfaction that I receive just makes me "fly."

However, despite my enthusiasm and joy, already I have more

invitations to present SWE than I can reasonably handle. My invitations, which come from all over the world, could easily keep me busy for the next ten years! So, I have come up with a plan that will help meet the demand.

My plan is to teach others to share SWE, so that all of us together can bring these frequencies to the world. By teaching others, we will be able to reach more people in a shorter period of time.

When I Visit

During my travels, I get the chance to observe first-hand what happens when people use the tapes. I play certain tapes so people can have the direct experience of the frequencies, and this way they know at least some reality of the power of the sound. During the lecture, some people tell me immediately what happened to them, while others share their testimonials with me at the end of my lecture. Each lecture that I give is really a blessing. It is rewarding for me to be able to do something I like, and to hear and see all the marvelous things that the sound frequencies do for people.

I was in Austin, Texas in October 1993, giving many lectures in Austin and the surrounding cities, when one day, Terry Spears, a wonderful lady I met at my Sacred Geometry class, phoned and invited me to go to her home town of Fairfield, Iowa to give a lecture. She had been using the frequency tapes for a month, and had begun to rent her tapes to others for $2.00 a day. She was so excited about them and she engendered excitement in those she told about the tapes.

I did not know that Fairfield, Iowa is the home of Maharishi University and that half the city's residents practice transcendental meditation. My friend had been doing TM for many years, yet she discovered that the frequency tapes were helping her to move through blockages in a very smooth and easy way. Many people in Fairfield had spent decades meditating daily, trying to reach a state of bliss. Most of them reported that TM had brought them many benefits, but they were still searching for that bliss state.

At first I was reluctant to think about traveling to Fairfield, I didn't really want to travel north in November. I was not used to driving in snow with a motor home. But my inner voice became louder and the message was very clear: I was to go! So I began to organize my trip so that it would include Fort Worth, Texas; Santa Fe, New Mexico; then on to Fairfield, Iowa.

Evelyn Beatty, a wonderful lady in Fort Worth, had experienced her SoulNote and the SWE tapes before. She invited me to give a lecture to her family and friends. So, with another friend, Kathy Rowlings, I embarked on this "mini-tour."

My next destination was Santa Fe, New Mexico, where two days of freezing temperatures gave me a taste of what to expect later. Here I did more speaking, met more people, heard more stories, had more fun—just what I had hoped for!

Next stop—Fairfield, Iowa. While I drove those hundreds of miles between Santa Fe and Fairfield, I had plenty of time to take a walk down memory lane and recall some important experiences and lessons. I didn't realize that once I got to Fairfield, I was going to be tested regarding how well I had learned those earlier lessons.

My "Pain in the Stomach" Lesson

As the road rolled smoothly beneath my wheels, I reminisced about my younger days in Canada. I had become involved in political action and was up-to-date about what was going on in government. I was disgusted by my findings. Nonetheless, after I moved to the USA and remarried, I decided I wanted to know more about the U.S. government. I read stories about the goings-on of our government that were incredible and deeply disturbing. I began to feel continuous discomfort in my stomach. I'd never previously had stomach problems, so I seriously tried to discover what caused my discomfort.

This happened in 1991 while I was still married to Rene. One day, I suggested to my husband that he read some information I had just received. When I asked him later what he thought about the information, he told me that he'd read the article in the bathroom and said "it was so bad it constipated me."

Immediately I realized that all the negative information I'd been reading about the U.S. government was causing my stomach problems. It had not dawned on me that as I was reading all those horrible stories, a lot of anger and other difficult emotions came up and this had created the pain in my stomach.

After I realized that my husband and I had similar experiences, I took all the books and papers, even some that had been very hard to get, and put them in the garbage. I didn't want to give them to anyone else because I didn't want others to experience the same kind of negativity. The garbage was the only place for them.

Since then, I have had no more stomach problems and I have refused to read anything of that nature.

Powerful Lesson in Pasadena

I then began to remember a time when my car broke down on a freeway in Pasadena, California, near Los Angeles. Eight lanes of cars were constantly moving on this freeway. When someone would just pull over to the side of the road, without actually blocking anything, the traffic would slow down a lot. I had experienced this many times, but always as an impatient spectator.

I was now the one creating the slowdown. I used the roadside emergency phone to call for help, then waited in the car.

As I sat there, I started becoming aware of a lot of negativity coming in my direction. I realized that I was being bombarded by countless peoples' feelings of impatience, anger and rage. Even though I was not blocking traffic where I was situated on the side of the road, I still was the object of rancor. People slowed down and sternly stared at me. I felt sorry that I was the cause of the slow traffic, but since I was there, and unable to go anywhere until someone came to rescue me, I decided to fully experience this situation.

Suddenly I had this big *aha!* It dawned on me that I was experiencing the reverse of the situation I had created when I was sending out negative thoughts to the people I was reading about in books about the government. This time I was the recipient of the negative thoughts instead of the initiator.

I realized that regardless of whether I send a negative thought or receive one, it creates the same destructive effect. I suffered stomach distress from sending out negative thoughts to non-specific government agents. However, here on the freeway, I was *directly* receiving thoughts of negativity and feelings from stressed drivers. This was *not* fun! But what a vivid lesson it was for me.

I realized that when we are on the freeway together like that, we depend upon each other, whether we know it or not. It is like we form a big river and each drop of water has to follow the flow. The bigger the freeway, the bigger the agreement with others. I realized that I need to have a lot of trust in others. I have to be ready to go with the flow. If I get impatient or angry, the person I hurt the most is me.

Since that time, I have simply learned to surrender when I am slowed down on a freeway, or by other situations that create delays in my life. I try to send thoughts of understanding and compassion to those "causing the traffic jams" in my life. I have discovered that each time a delay happens, something better occurs than what I had planned. It's all part of a plan. I used to be angry at myself when I "missed" an exit, or took a "wrong" turn. Now I am more aware that these "mistakes" were my way of avoiding a real problem. "Going with the flow" became my motto.

Learning to Love Myself Unconditionally —
Light and Shadow Sides

While waiting in the car beside the freeway that day, I determined to put my "pause time" to good use by purposefully practicing loving myself—*all* of myself. I decided the first step to take would be to acknowledge and accept everything about myself. I knew until I could do that, I couldn't feel unconditional love for myself.

As I sat there with the cars and drivers storming past me, I took a mental inventory. I acknowledged all the positive and negative parts of myself. I knew that avoiding or denying the shadow side of myself would not make it disappear. I had no problem accepting the Light of day and the Dark of night. So I proceeded to apply that acceptance to myself and decided to change what I could and accept what I couldn't.

After practicing this for a time, I began to honestly love both the Light and Dark sides of Nicole. I then realized that what I wanted to do was to bring more Light and Love to our planet. I think the role I volunteered to play this lifetime is of a Light bearer and I want to play my role as best I can.

Naturally, I recognize that some people have volunteered to play the role of the dark side or villains. I appreciate and thank them for accepting this role since it is not a role I would like to play this time on planet earth. As far as people who are Dark bearers are concerned, I accept that the best I can do is avoid them, protect myself with Light and project good feelings to them.

Since this freeway incident, I have focused on projecting good feelings to everyone including those I see on the news. I decided to try to watch the news on TV once a week. It was a big commitment for me because I hadn't watched news for more than ten years.

When someone was arrested for murder, for instance, it dawned on me the bath of negativity the person received from millions of people all at once. I thought to myself that if this person had his heart open, the negativity would be enough to kill him instantaneously.

I noticed that I certainly preferred receiving feelings of love from people rather than feelings of hatred, so I began giving to others those same good feelings I preferred. It is literally true for me now that "I reap what I sow" and "what goes around comes around."

Fairfield, Iowa

The drive from Santa Fe to Fairfield proved to be a pleasant experience since it gave me an opportunity to reflect on such pivotal times, like the freeway incident. When I reached Fairfield, I parked in a beautiful back yard where hundreds of ducks came to eat and swim in the pond. In addition, there were peacocks and peahens and I felt like I was in a beautiful private park.

The first evening in Fairfield, Kathy Rowlings and I presented our respective works. A lot of people were interested in having their SoulNote done and also in using the tapes. Many had already rented a tape from my hostess Terry Spears, so they already knew what to expect.

Request and Discovery

The atmosphere in Fairfield, Iowa, was certainly conducive to spiritual growth, and there was very strong interest from the people in Fairfield regarding the tapes. Their strong interest caused me to learn more about my own work.

Some people in the Fairfield community meditate four hours a day and work part time. They could not play the tapes when they were working and yet some were so excited about the results they had already experienced, that they wanted to know how many tapes could be listened to simultaneously. This was important to them since the only time they could play the tapes was during the night.

I did not know the answer and so I consulted with my guidance. I was told that it would work well to listen to three tapes at the same time. More than three would confuse the body. People eagerly responded to this information and they searched stores in order to find a tape deck that could play six tapes consecutively. Many meditators purchased three of these tape decks and were able to play the tapes all through the night. In six-and-a-half hours, they could play eighteen tapes, since all tapes are one hour long except for the **Chakra #3** tape, which is ninety minutes in length. Of course, I now have twenty-five tapes available [in 1994; as of 1998 edition, forty-seven recordings are available - Ed.], but obviously this same approach could be used regardless of the number.

Love, Light & Protection

I had a very interesting experience a few days after I arrived in Fairfield. Since the beginning of my journey, I had parked in many RV camps. I learned to lock the door of the motor home when I went to bed or left the park area. But when I got to Fairfield, Iowa, where 50% of the 10,000 inhabitants meditate, I decided it was unnecessary to lock the door.

Kathy slept in our hostess' house during our stay. When she would wake up each morning, she would come into my motor home and prepare coffee as part of her regular routine.

One morning, I was going over my dream to be sure I didn't forget anything, when I heard somebody open the door to the motor home and enter. Thinking it was Kathy in search of coffee, I continued reviewing the dream. I heard someone go outside and then inside again then outside, and I noticed that no fragrant aroma of coffee was wafting my way.

Eventually, I decided to get up and go into the house. Because of the cold weather, I didn't want to use the septic tank in the motor home, so I had accepted my hostess' invitation to shower in her house. When I got inside, I discovered that Kathy was still sleeping. As I arrived, she awakened. I asked her if she had just come into the motor home. She said she had not. I told her that somebody had, but it must have been an "angel."

Within a couple of minutes, the doorbell rang. Our host and hostess were still sleeping so I answered the door. To my surprise, the Fairfield sheriff greeted me. He asked if we had seen a man driving a white pickup. Immediately, I realized what had happened and asked the sheriff to wait.

I returned to my motor home and found my purse on the couch just in front of the door. My wallet and the $700.00 in cash that it had contained were gone. In addition, all my credit cards, my green card and my driver's license were gone.

I instantly recognized this to be a spiritual test. My natural state is to trust people, and my life is generally quite harmonious and flowing. I realized that this was a test of how well I could apply some of the lessons I had been practicing and reflecting on during the drive to Iowa.

I closed my eyes and sent the man who had taken my wallet thoughts of understanding and love. I mentally told him that I didn't have any bad feelings toward him, in fact, I wanted no connection whatsoever. I projected the thought that I would agree to give him the cash, and that I hoped he would use it to eat and find a job. I then mentally asked him to return all my cards and papers. This took a couple of minutes. I then returned to the house where the sheriff was talking with Kathy. I told him about the entry in the motor home that I'd only heard. I reported to him what was missing and he left.

Two hours later the phone rang. Serendipity was at work!

Someone had found the wallet and had discovered the check in it that my client had written me. The wallet finder called my client and asked where I could be reached in order to return it all to me. The client had had a session with me just the previous day and was able to tell the finder who and where I was. The wallet had been found in the middle of a street ten miles south. The wonderful finder called me and I went to pick up my wallet. Only the $700 cash was missing, all else was returned to me.

Seven hundred dollars was quite a bit of money for me to lose at that time. Yet I didn't want to create any more guilt for this man nor did I want any resentment or hatred inside myself. I thought that by "giving" him the money, he may feel good and reflect on what he had done. Who knows, maybe he would never do it again. But the main lesson of the incident was for me. To act this way was an act of love for *me*, even more than for him. I kept my peace of mind which creates happiness for me and I continue to trust people and to leave not only my door open, but my heart as well.

A Dream of Non-judgement

My experience with the man who took my wallet gave me the opportunity to put into practice a number of spiritual beliefs, not the least of which was non-judgement. Through the years, I began to realize the power of the mind, and along with this recognition I became aware of the power of judgement. If I had to choose which I believe is more negative—actions or judgement, I think that for me, judgement would be worse.

When I judge someone or something, it inevitably sets up a scenario where I think I am better in some way. I realize that, like most of us human beings, I have spent my life judging others and I think this definitely creates negativity. I have unquestionably done my share of judging and therefore creating negativity.

When this dawned on me, I stopped to wonder what I could do to reverse the pattern. I decided that the best way to begin would be to forgive myself for all the judgements I had originated and to also forgive anyone who I felt had judged me. This seems to me the only way to develop unconditional love for myself and others.

Judgement is a tricky thing and sometimes sneaks up unexpectedly, but I dream of the day when I will not have an ounce of judgement left in my heart. When I achieve this, I really will have created Heaven on Earth all the time.

I did a great deal of reflective thinking while I was in Fairfield, Iowa for five weeks. I met wonderful people and was constantly busy testing people for their SoulNote and producing tapes. My host and hostess offered to videotape my presentation and this video has become a useful tool for me. In fact, I still use the video that they taped to this day to introduce people to SWE tapes.

I was grateful for Terry's hospitality. It was such a pleasure being in Fairfield. Before I left town, Terry organized a meeting so that all of the people that I had worked so closely with while I was in town could come and ask questions. Twenty-seven people came to the meeting and many shared their responses to the tapes. A number of people noted a feeling of being in a state of bliss when they played the tapes. These are the kinds of testimonials I love the most because they imply so much.

Crystal Lady Alora

Terry had many huge crystals, called Earth Keepers, in her home in Fairfield. She had placed a classified ad in a newspaper looking for a knowledgeable crystal expert who could help in the activation of these enormous crystals. During the last week I was in Fairfield, a lady from Europe named Alora answered the ad and said she would travel from Delaware, where she was at the moment, to Fairfield to do the activation.

My work in Fairfield felt complete and I was ready to move on except I intuitively felt I had one more person I needed to meet before I departed. As a result, I waited to see who the Universe had in store for me.

In retrospect, it seems obvious that this "one more person" was Alora. Terry and I went to meet Alora at the bus station. Terry was looking forward to meeting Alora because of her ability to talk to crystals and ask them to awaken—thus activating them. When she arrived at the bus station, we made our introductions and began to talk. In no time at all she was sharing with us that she had not been sure she should come, but a psychic had told her she needed to meet a lady who was from New Mexico and it was important for her to come because this lady would be in Fairfield at the time of Alora's visit. It didn't take us long to realize that I was the "lady from New Mexico" that she was supposed to meet and that she was the "one more person" I was waiting to meet.

By now it was December and I decided I would travel to the East Coast to give a couple of lectures and then go to Montreal to spend a couple of weeks with my children for the holidays. My plan was to be on the East Coast by the following Saturday evening. I learned that Alora needed to be there Saturday morning. So I decided to leave a half-day earlier to accommodate her schedule and thus she became my traveling companion.

We had a delightful time sharing thoughts, philosophies and ideas and the first day of the trip was ideal. But on the second day, a big snow storm kicked up and threatened to impede our progress. The visibility was nil and cars and trucks were off to the side of the road everywhere we looked. The road conditions were terribly slippery and we still had 500 miles to go.

If we were to arrive on time, we simply could not stop for the storm. I talked with my higher guidance and asked for help because Alora and I felt that we really had to continue.

Less than sixty seconds after having asked for help, a sand spreading truck pulled out in front of us and I was able to follow it almost all the way! The truck made the road safe for us to travel and we drove through the storm like we were charmed—which we were! We safely arrived at our destination on time.

The only "problem" was that the motor home and my maroon car-in-tow, Toro the Taurus, were covered with mud. Every square inch of both vehicles was mud-covered. What a mess!

Alora had proven to be a wonderful traveling companion. She told me of her work with crystals as we amicably chatted the cold miles away. When we got to Delaware, she introduced me to several people she knew there. Later she also opened some doors for me in Hawaii as well. Alora was part of the divine plan—we were to come together at the perfect time under the perfect circumstances for both our highest goods. She was one of the many wonderful people that flowed through my life while I was on the road.

Within a week, I was able to see the people I'd come to see in Delaware as well as Alora's friends. I implemented my plan to continue driving to New Hampshire to visit a friend before traveling on to Montreal. I vividly remembered the mud bath my vehicles took getting to Delaware so this time I asked for a "clean" trip to New Hampshire.

Universe complied perfectly. What I got was rain, rain and more rain. There was so much rainfall that I had to stop my car many times due to zero visibility. At least it cleaned the motor home and tow car! I thought I had learned to be precise when I asked for something from my guidance, but I realized I needed to re-word my request. For the final leg of the trip from New Hampshire to Montreal I requested a dry and sunny trip. I got it, but it was soooo cold. I'm still practicing my "purchase orders" for weather.

Back Home in Montreal

My son Francois had reserved a place for me to park the motor home once I arrived in Montreal. I had brought a full propane tank with me so the little electric heater I had would heat the motor home. I was prepared for any kind of weather. I was assured that, because of the heat, I would have no problem with water freezing in the motor home's pipes and so I felt comfortable leaving the motor home while I was gone for our Christmas get-together.

The next day, I left Montreal in my Taurus with my two sons, my daughter-in-law and my two cats. We had rented a condo at the Laurentides. It was a perfectly beautiful place to spend Christmas. Soon after we arrived, the temperatures took a nose drive and became so cold that no car would start.

None of us worried because we had all the food we needed, a beautiful view, and we were all together. We had a wonderful time playing indoor games and enjoying being together. We had no doubt the weather would return to normal soon.

The frigid temperatures persisted throughout the end of the week. When we tried to start the car, it would not make a sound. It was dead, dead, dead. We called a garage to get help only to discover that it had been inundated with similar requests and had a two to three day backlog of frozen cars ahead of mine. Clearly the only recourse was a miracle.

A Toro Miracle

My poor car Toro the Taurus was used to the warm, dry climate of California, and other than adding antifreeze, I did little to prepare it for the icy temperatures. I could not blame it for not wanting to start. However, we needed to go. Our lease was up on the condo and so it was time to go back to Montreal.

I asked the Universe for assistance. After blessing the car, I explained that we needed Toro to start so we could depart. I went to start the car, fully relying on a miracle. And guess what? Toro just started right up! So we loaded the car, including an extra cat that had found us while searching for food and warmth, and returned to Montreal.

My sons had planned to visit their father for New Years and I decided to surprise my family—brothers and sisters who still lived in Alma approximately 300 miles from Montreal. Toro faithfully enabled me to arrive on New Year's Eve around 9 p.m. My family was at my sister's home eating dinner when I arrived. The first thing they did was give me a sermon because I had driven 300 miles across "The Park" and they feared for my safety. The Park is a national park area which is deserted. If a person were to get stuck, there would be a long wait to be rescued. After the mini-lecture, they asked me what I was doing in my life. I had not given them much news of myself during the past year, so I began to share my work with them. I was eager to let them know about all of the good things the SWE tapes do for people.

"A Prophet. . . In his Own Country"

I was accorded my family's attention for fifteen minutes, but after that they resumed their talk of friends and family members. The words focused on this disease or that, of the ones who had died or were close to death, etc. My work was of little interest to them and I realized the truth of the old adage about "a prophet is not without honor, save in his own country."

This was the very first time since I had begun this work that I felt it was not understood.

I almost felt hurt because I knew that the frequencies could help them resolve their problems. But my family did not want to even know about the recordings, let alone use them. I felt more sad for them than for myself, I would love to have seen them more interested. But they weren't—and that was okay. My job is not to change them, but to love them exactly as they are.

When they asked me when I would return, I answered "in fifteen years." I didn't want to hurt them, of course, but I realized that it doesn't serve anybody to push or force something on them. So many people can be helped—people who sincerely ask for and are interested in this technology—that it is my desire to serve them and simply love all others, including my own family.

This experience with my family, however, helped me to understand people who come to me to get recordings with the intention of helping a family member. Their intentions are good, but the majority of the time, it doesn't work. The actual users need to make the decision and exert the effort on their own behalf in order to achieve the full benefit. This of course does not apply to a child or an elderly person who is under someone else's care.

Live and Learn. . . and Learn and Live

All my life, I have experienced a great deal of pleasure when I have been able to give something to someone. Because I am a giver by nature, and feel so strongly about the benefits of the SWE tapes, I would often give them to people when I first began this work. I was especially eager to give the tapes to people for whom I knew they could make a big difference.

Early on, if a person told me they didn't have the money to buy the tapes, but would like to have them, I would just give them away. And I felt good about doing so. But after a while I discovered that almost all people to whom I gave tapes were not using them. Naturally this meant they also received no benefits from them.

In this way I learned first hand the truth of the saying that "Something received for nothing is often valued as nothing." As a result, I have changed my policy. I don't give away the tapes any more. I've made this shift because I really want people to have optimum results. Even if it pleased me to give away a tape, I didn't really render good service to that individual. I am here to serve, not to please myself!

My Frozen Motor Home

When we returned to Montreal, naturally I checked in on my motor home. Here was a beautiful example of the best laid plans going awry. It had totally frozen during the sub-zero cold snap. My electrical hookup supply had been shut off while I was gone and the water heater was broken. Despite the precautions I took, my motor home succumbed to the cold weather. But my son Francois helped me take it to the garage for repair and soon it was roadworthy again.

I thoroughly enjoyed the time I spent with my children—it was a great blessing. In addition, I produced more SWE tapes so I could build up my inventory in order to be ready to take to the road again. I hugged my goodbyes in bone-chilling 40-below-zero temperatures and headed south.

The Sunshine State

I'd had enough cold for one winter! I discovered that a motor home is not the ideal place to be when it is 40 degrees below zero outside. So I headed south with enthusiasm and looked forward to more agreeable temperatures

On my way south, I stopped in Delaware to reconnect with the wonderful people Alora had introduced me to. A couple of minutes after I arrived at their house, the phone rang and Doris, a lady who'd been using the tapes during the last month, was calling because she was upset.

On one hand, she had a wonderful story about how she began hearing music in her head after listening to the tapes. I was delighted to hear that she had begun hearing music in her head as a result of the frequencies.

But on the other hand, she was not happy with the feelings she started experiencing in her second chakra. At seventy-two years old she didn't know what to do with all the sexual energy she had. These energies had been repressed after she lost her husband over thirty years earlier.

After we talked, she realized that by allowing her feelings and transmuting them instead of repressing them, she had opened a chakra that had been closed. By doing so, the energy was flowing freely into her body. This was what allowed the music to be heard.

Taboos from Brainwashing

How many of us believe that having a sensation in the genital area is almost something we have to be ashamed of? I remember my parents viewpoint that sexual relations were supposed to have only one purpose: making babies. No wonder families with fifteen or twenty children were common. Such ideas meant if you passed the age of making babies and had feelings in the genital area, the first thing that came to mind was to shut it off.

The feelings don't have to be consummated with a partner. It is perfectly all right to acknowledge these feelings and to transmute them by raising the energy for higher purpose. But we have to pay attention to all our feelings whatever they may be: pleasure, pain, guilt, fear, etc. . . and deal with them. This is the way we choose the more beneficial feelings for our greatest good. We incarnate in a body to have a chance to experience all the emotions. When we deny them, we simply delay our spiritual growth as well as create physical problems.

There is tremendous energy located in the second chakra. It is not an accident that many religions have perverted the truth about the sex issue and created so many rules, based on fear, to prevent people from using this great source of energy. All the taboos around sex cause the extreme abuse of it; it is done in "secret" and becomes even worse when we feel the need to hide it. It eventually becomes an obsession. We all have experienced the reaction to something that is forbidden. Sex is a beautiful gift when we use it in a balanced way. It is like food. We need some food to stay alive; too much can make us sick, no food leads to death. Balance is always the key.

I stayed in Delaware for a couple of days to work with some more people and made some more nice acquaintances. The first morning that the rain started pouring, I decided it was time to continue my journey south. Soon after I left Delaware, the weather cleared up and I eagerly pointed in the direction of Florida.

I reached the Sunshine State free of trepidation despite the warnings on news broadcasts that people driving RV's were being attacked and robbed on the freeway going into Florida. I was determined not to let anything prevent my visit from happening, so I drove on. I was very confident that everything would be perfect, and it was. I arrived in Gainesville without a problem.

Jeannette Edwards, a wonderful lady I had met in Fairfield, was waiting for me in Gainesville. She lived by a beautiful lake. So there I was, at the end of January, basking in sunshine and thawing out from the Canadian winter of a few days before. I parked at her home and she organized a lecture there. I worked with quite a few people and made some nice new acquaintances as I spent time in Gainesville. I didn't know at this time that Jeannette would be coming to Colorado to help me with my work. But I surely sensed that she was a very special lady.

I drove to Orlando and spent a week in an RV park until I accepted Dr. Roy Kupsinel's invitation to stay in his back yard. While there, I had the opportunity to go on an Intuition Cruise and travel the Caribbean on the biggest ship afloat. And although I had just two day's notice, I felt comfortable giving lectures and working with people on the cruise. After returning to dry land, I went to West Palm Beach, Lantana, Stuart, Fort Lauderdale and back to Oviedo, Florida.

Everywhere I went I was able to test people for their SoulNote, provide SWE tapes and give lectures about my work. I continued to meet many wonderful people who were very excited about the frequency tapes and how they could be beneficial.

A Reminder of the Density

After I left Oviedo, Florida, I stayed in a beautiful state park camping site in West Palm Beach. Since there wasn't a sewage receptacle at each slip, I had to go to the dump station before I departed in order to drain the sewage.

I drove to the spot and then adjusted the tube in order to dump the sewage. I didn't realize that a little attachment on the nozzle had broken. As a result, when I opened the dump valve, the tube didn't hold and I got my hands filled with excrement! Ah, the adventures of traveling with a motor home, I laughed.

I was reminded that I had a choice. I could be mad at myself, embarrassed, or laugh. I recalled a time thirty years ago when I fell on the ice and got my pretty coat all wet and muddy. A bus load of people had observed this event and the people on the bus laughed loudly at me as I took my spill. I felt totally humiliated.

Later in life, I discovered that laughing at myself helped to transform a feeling of humiliation into levity and lightness. I noticed that I ended up laughing later at almost anything that I had experienced because I was able to see it from a new perspective. So, I reasoned that if something seemed embarrassing at the time, I knew I would be laughing at it in six months so why not just get on with it and laugh right now. Why wait! Because I learned how to release the event immediately through laughter, I was able to avoid being hurt inside. This approach has lightened my load many times. And this was one of those times.

I had found also that when I began to accept myself, mistakes and all, I could accept others more easily. I became less judgmental of others and I noticed that judging others decreased in direct proportion to the increase in my ability to laugh at myself. I also know that invariably there is a gift for me inside a challenge, if I look for it. So now I try to be grateful immediately when confronted by a challenge. I have faith that if I cannot see the gift at that exact moment, I will be able to see it later.

With these thoughts in mind, I cleaned up my sewage spill and myself and returned to my site. I picked up the cats and drove toward Oviedo.

I thought that my adventures for the day were done, but no more than fifty miles down the road, a giant rain storm started pouring so hard I couldn't see. The roads were flooding and the wipers couldn't go fast enough to keep my windshield clear enough for me to see. I had to slow down to let the big storm pass over. I'll never know what that rainstorm saved me from—but I'm grateful for it.

Smoke Gets In My Eyes

A few miles after stopping for the rainstorm, I suddenly realized that I couldn't see *inside* the motor home. I also became very aware of a strange smell. The inside of the motor home filled with "fog." Rain continued to pour outside and the fog inside became thicker. I had to get to a garage fast! Fortunately, an exit appeared and I took it.

I arrived in a little town and found that it had a garage. In order to see better inside, I opened my door and became instantly drenched. When I could finally see clearly again, I drove to the garage entrance. The mechanic discovered that a hose that runs from the radiator to the hot water tank under the sink had been severed. He was able to replace it quickly. By the time he finished, the rain had stopped and I was back on the road, thinking about all the interesting events that had happened since the morning began. I'd experienced solidity, liquidity, and finally vapor. What a trip!

New Self Discoveries

Back in Oviedo, I was welcomed by Roy Kupsinel, who once again invited me to park in his peaceful back yard. People who had experienced the SoulNote and the SWE tapes wanted their friends to experience them, too, so word spread rapidly. I quickly became so busy I had to back-order my tapes as I could not create them fast enough and still keep up with my speaking schedule and SoulNote appointments.

One day, while waiting in Roy's utility room, I decided to stand on the scales. What a shock stared me in the eye! Since I had started this journey, I had gained more than twenty pounds. My elastic-waist clothes had "protected" me from this awareness

The most startling part of this whole saga is that prior to using the SWE tapes, my back became painful if I gained more than three pounds! Yet here I was with an extra twenty packed on and no back pain in sight!

I was extremely pleased to learn that my previously fragile bones had become so strong that they could carry more weight without any problems. And although I was not thrilled about my weight gain, I was joyful over the obvious benefits the tapes had provided me.

I thought about the weight gain for a moment and realized that I had over-eaten, not because I was hungry but because I was eating sociably. I had eaten sociably with many people throughout this vagabond phase.

The good news was: I didn't need an x-ray to prove to me that twenty-five years of taking supplements had resolved nothing relative to my osteoporosis, but the frequencies had.

Robert to the Rescue

I had gotten extremely busy and was at a point where I needed help. I was very busy with appointments, but I had a tremendous backlog of orders which continued to build.

One day my son Robert called me and said he'd like to come to be with me in Florida, and at the same time he offered his help. I was absolutely delighted that Robert wanted to come down to Florida and work with me. I could not find a better person to take care of the production of the tapes. Robert is very focused when he performs a task. I long ago discovered that it is important to have a good attitude when working with the tapes in order to be extremely accurate and produce tapes of the highest calibre. He was "good attitude" personified.

Off to Colorado

It was a blessing to have Robert with me. He loves this work and he enjoys the benefits of the tapes because he is immersed in all these wonderful frequencies since he reproduces them. My staffing situation was ideal.

At the beginning of May 1994, I returned to give another lecture in Gainesville and then went back to Oviedo, Florida. Once there I received information from my guidance that I should prepare to move to Colorado. I was really pretty disappointed because I had promised some people that I would go to their towns and I had not yet done that. I satisfied my momentary surprise by reminding myself that I could come back later after getting settled in Colorado.

The purpose of moving to Colorado was to find land where we could put an Awareness Center of Sound and Light, a Research and Conference Center, a Wellness Center and establish a community. It also served the purpose of helping us get situated in a larger home. The lack of space with two of us living in a motor home, me testing people and Robert producing the tapes, was challenging .

By mid-May, Robert and I were on the road to Colorado. It took us only five days to arrive in Santa Fe, NM, and it was a very enjoyable trip.

We got all the way to Santa Fe without knowing where our exact destination was to be. I was waiting for more guidance in order to know exactly where to go in Colorado. At the last minute, I was guided to go toward Durango. My friend Kathy Rowlings, whom I had met in California, was already living there. She had moved to the Durango area after I had told her how wonderful it was, but she had no idea that I was about to join her in the vicinity.

I contacted Kathy and told her I was guided to come to Durango. She was thrilled and immediately began a search for a rental house for Robert and me. She found a little house in the Bayfield area about eighteen miles east of Durango. It was situated beside a creek and the property possessed a little pond as well.

Rentals were difficult to find and so Robert and I decided to move into the little log house. What the house lacked in attributes was more than compensated for by the lush surroundings. Deer roamed our back yard and a beautiful mountain stood majestically at the end of the meadow which served as the "front yard." Tall evergreens provided abundant shade in the summer. Hundreds of birds, not the least of which is the wonderful Colorado state bird, the hummingbird, added to the beauty.

I instantly enjoyed the Colorado climate and weather. I had spent ten years in sunny states like California and Florida, but this was more than their equal. I discovered that the sun shines almost every day and the humidity remains comfortably low. Even when the temperature reaches 80 to 85 degrees in the summer, the air is very comfortable and it's just perfect for me. Robert also loves it here. He didn't care for California or Florida, but Colorado is probably becoming more his place than Quebec, a place he has always cherished.

I also discovered that the deer loved my roses as much as I did. But it was worth letting them munch my blooms just to get the opportunity to observe them. When they'd come into the yard, I couldn't get enough of their presence. I admired their elegance, gentleness, agility and purity. I would love to hug them. I developed a deep love for deer. Interestingly, Maud Sejournant, this wonderful lady that does shamanic work, guided me into a meditation where one finds her animal power, and one of mine is the deer.

I have a knowing that the time is not too far off when we humans will be able to have "wild" animals around us and live in complete harmony. Animals have so much to give and so much to teach us. I dream of the day when I will stay in a location long enough to make real friends with deer and other wild animals.

Finding Our Awareness Center

Keeping my Florida Promise

I visited Florida twice after moving to Bayfield, Colorado. On one of these trips I stayed with my friend Lynn Veitch in Winter Park. One morning we heard a cardinal sing and this prompted Lynn to go outside. She carried bird seed in her hand and quietly talked to the cardinal and invited him to come and enjoy the treat. In time, the bird developed a trust for her and ate out of her hand.

As I watched this scene, tears of joy came to my eyes. I realized that both Lynn's and my connection with the animal kingdom was becoming stronger every moment.

The very next day, Lynn was busy with a client when I heard the cardinal sing. I found the bird seed and went outside with it in my hand. After a couple of calls from me, the cardinal landed on my hand to eat his breakfast. What a treat for me! I know there are already many who the birds trust this profoundly, and soon there will be millions of us enjoying harmony with all the kingdoms.

Communicating with Insects

Lynn recommended that I read *Kinship with All Life* by J. Allen Boone. In this book the author recounts many wonderful animal stories and compellingly and vividly describes his own very real communication with animals as well as flies and insects. One of his stories reminded me of an experience with ants I'd had when I moved in our new house in Vista, California, years before.

We were being overrun by millions of ants coming into our house from outside. They would cross the kitchen ceiling to reach the other wall and then return outside. Our ceiling was like a ten-lane freeway with five lanes going each direction.

For the life of me, I couldn't see what was attracting them. Previously, I'd used spray to get rid of them. It would work for a while, then they would return. Moving in a new house, I didn't feel I wanted to kill anything and decided to talk to them and see if I could make them understand.

I told the ants that they were not invited inside our house. I told them I knew there was plenty of food outside on all the nearby fruit trees. I also told them that they would have to leave the kitchen or I would have to destroy them. I gave them one hour to leave. I didn't do anything else special, except talk aloud to them *with intention!*

After an hour I returned to the kitchen. The ants were still there, but they were in a special formation at the top of the ceiling. They had formed a circle about three inches in diameter. I guessed that they were in a conference, so I addressed my request to the one I thought was in charge of the group. I spoke to him out loud and said I would give them another hour, but if they remained in the house after that, they would be sprayed and killed. Another hour passed and I returned. This time they were gone; very few were still there. Finally they all left and never came back into the kitchen. They came into my office a couple of months later but since there was no food there, they eventually left. Maybe they were attracted by the sound wave tapes!

The St. Francis of "Flies"

During the four months we lived in the Bayfield house, Robert and I learned to talk to flies. We discovered we could effectively ask them not to land on the food or on our bodies. We were surprised and happy that they responded to our requests.

Also during our time in Bayfield, we noticed that house plants loved the SWE frequencies. My geraniums and Christmas cactus were continuously in bloom. The foliage plants grew more quickly and beautifully without any other special care. I cannot wait to have a garden where we can use sound to make all the fruits and vegetables grow larger and healthier.

Flying Business Trips

While living in Bayfield, I made business trips to Boulder and Denver, as well as much longer sojourns to Miami and Orlando, Florida and three Hawaiian islands.

Naturally, flying to Florida was much speedier than driving in my RV, but I had to carry my heavy computer and speakers with me. Before, I'd been used to traveling in my motor home with the electronic equipment right there with me. Believe me, I can't wait until I can get a laptop that will accommodate my needs!

A few times the subwoofer speaker in my luggage was damaged in transit, but somehow I always managed to get things fixed and taken care of so that SWE work could be shared.

When I wasn't traveling and was at home, people would come from many states as well as Canada. One even came from Europe. They all came to have their SoulNote done. If they had difficulty adjusting to the 8,000 foot elevation, I would play the Prana tape continually, to help them breathe more deeply and take in more oxygen from the thinner air—so they could sleep well.

My Dreams are Answered

I took some time to become acquainted with Southwest Colorado since I was looking for land to buy. My lease in Bayfield would be up by the end of September 1994, and so I figured I'd better start looking.

It was love at first sight when I discovered Pagosa Springs, Colorado. I marveled at the spectacular views all around me. I phoned a real estate agency and inquired about leasing a house in the Pagosa area. It seemed to me that a number of people were moving into the area and so I didn't know if I would be able to find a large enough home, since demand for housing in the area was high. I very much wanted to go on with creating a center where people could come from all over to have their SoulNote done, and also give them a taste of what heaven looks and feels like.

Results happened effortlessly. The very next day, Robert received a phone call which alerted me that a house "may" become available. Two days later I visited it and loved it. It suddenly had become available and so I signed the lease agreement and contentedly told myself it was perfect for now.

I was very grateful to have found this house so easily. When things that I ask for manifest quickly, like this house did, I know that I am "in the flow." When things don't come quickly, it means one of two things: either that I am not ready and the timing is off or it is not for my highest good, despite appearances to the contrary.

I had always dreamed of living on top of a mountain. I'd already enjoyed living in a beautiful house half way up the mountain. But the view it offered was nothing like this one. Here we enjoy nearly a 360 degree view of the mountains. It is spectacular! People in Pagosa Springs tell me that it is one of the best views around. It is located less than three miles from town, but it is like being in the country. The wonderful thing is that I also have the facilities of Pagosa Springs nearby.

Here we enjoy the four seasons and the sky that constantly changes. Sunrise each morning casts different pastel colors on the peaks—peaks which are almost always snow-covered. Sunset brings its beautiful color palette to the mountain top as well. I am truly in heaven thanks to all this continuous beauty.

The various pools and spas of the hot springs are probably more of a meeting place than the local restaurants. A lot of business is conducted in a bathing suit sitting in the mineral waters.

There is nothing fancy about downtown Pagosa Springs. However, the beauty of the surrounding area, here in the Rocky Mountains, is just breathtaking. City lights do not obstruct the view of the night sky and humidity is low and the air pollution is minimal. The daytime sky is a vivid blue well over three hundred glorious days each year. There is a mix of mountains, hills and meadows. Many herds of elk migrate through the valley below our house in fall and spring.

We moved into this ideal setting at the end of September 1994. Just three days later I found myself flying to Denver for a three-day conference and to New Hampshire for a symposium on sound. As much as I was interested in the symposium, I couldn't wait to get back to Pagosa Springs to admire Mother Nature's fantastic fall spectacle. Fortunately, I was back in time to enjoy the first snowfall in mid-October.

It was about this time that Jeannette Edwards from Gainesville decided to come to Pagosa for a visit to see if she would like to live here. After coming for an initial first visit, she returned in December to spend four months with us. I was delighted to have her here. Jeannette's experience in hospital administration and her excellent English made her the perfect person to create with me the business plan for Sound Wave Energy. In December, I went on another Intuition Cruise and Jeannette and Robert stayed here to take care of SWE.

The Book - *Just do it*

When I began to travel in connection with my work, I got a clear understanding that I needed to write a book about my work. A book would be helpful both in introducing the tapes to people as well as providing a long-term, in depth resource for people to refer to. I knew the idea of having a book made good sense. However, I just could not see myself writing a book in English—after all, my native tongue is French.

It became very clear to me that I was the one who had to sit down and put the words on paper. It was my job. I had a number of good excuses why I didn't want to do this, but my inner guidance told me clearly that it was the right approach. I committed myself to "just do it," as they say at Nike.

From My Heart to Your Heart

I wrestled with a number of "*who me?*" doubts—including the fear that the computer would "eat" my work. I had been asked to write an article on the subject of Love for *Aurora Rising* Magazine. I began writing the article on my computer, but after many days' work, and eight pages of composition, my computer froze and I lost half of my work. Obviously this experience did not bode well for the much more extensive work that I knew would be involved in writing a book! Nonetheless, I persevered.

I decided to organize the book along the same lines as my lectures. When I first began lecturing about the work, I tried to explain a lot of details about the frequencies and focus on a great deal of left brain stuff. The technical information was not the most exciting aspect of the work except for a minority of people attending the lecture. The majority were more interested in the experience of the sound and what the sound could do for them.

I wanted to find a middle ground where I could connect with everybody—both the right and the left brained. I realized that literally thousands of people had told me after lectures that they loved the stories I would tell about the tapes because they could relate to them. This caused me to change my presentation somewhat and begin to tell stories about SWE tapes and how they had helped me and other people as well.

This approach really improved my audience attention level. I discovered that people really liked hearing the stories, and it was also what I felt most comfortable with. So I decided the story approach worked the best to achieve my goal when I was speaking. The same approach, I reasoned, would be effective for the book: stories were the theme of the day!

As a result, the plan for the book was that I would tell my life story. And in order to accurately portray my intentions, facts and feelings in the right chronological order, I had to sit down and write myself. I didn't want my work to be represented by solely intellectual aspects. It is not my diploma that brought me to this work, it is my life. So, I realized it was a simple task of writing from my heart to your heart. Just true stories, shared in a simple way, would convey the important message, I hoped, "if I can do it, you can do it, too."

During the first three months of writing the book, I told my guidance that it had my permission to wake me up when it was appropriate for me to write. As a result, I woke up at all kinds of hours like 2:00 a.m., 3:00 a.m. and 5:00 a.m. My personal habit has been to sleep very well and not wake up during the night at all. So when I became wide awake in the middle of the night, I knew it was time to get up and sit at the computer and write. Six or seven hours later, I would complete my writing for that time, take a shower and go downstairs to have coffee or hot tea and go about my day.

By now it was the end of March 1995, and not only was I writing the book, but I was guided to put together the frequencies for a new series of recordings.

I had been given these frequencies two years earlier, but had not been allowed to release them. Every month for two years I would ask my guidance if I could go ahead and produce the tapes. The answer was always "no," until one night I suddenly I received a "yes" answer. It was time to produce the tapes. I had anxiously awaited this event and now here it was plunked in the middle of my book-writing project after all this time. Nonetheless, the green light was on and so I wanted to create these new tapes.

As a result of the new series, as well as the creation of the newsletter that I had started in June of 1994 with Roy Kupsinel, I was kept very busy and put the book on the back burner for awhile.

The newsletter had been published twice in 1994, and I did not want to publish a third edition without including the information about the new tapes and the fact that those tapes were ready. My policy has been that anybody who buys even one SWE tape or CD gets put on the mailing list to receive free newsletters. These news-letters are to keep people informed of what's going on, including the development of the center.

But the book obviously needed to go on the back burner for a bit while Divine timing was having its way with the project.

Angels are Sent

I finally sat myself down and began writing the book. It had become quite clear that this was what I was being guided to do—but after that, I didn't know what was going to happen. I only knew I had to start the whole thing myself and write down my story. After that, it was total trust.

I just love the way the universe works. It really does handle the details. So while I was doing my part, someone I had met in a far off state was about to come into my life. Something was happening!

Jill Lawrence, a journalist living in northeast Ohio, was having a kinesiology session with a woman in Akron, who introduced the SWE tapes to her. Jill called one day to place her order for the SWE tapes, and then called a few subsequent times with questions (she is an avid user of all tapes and says they're one of her most treasured possessions). When Jill and I began to talk, things finally started moving in the direction of manifestation of the actual book. It seems that the Universe sent me Jill to help me complete the work!

We quickly evolved our working pattern: I had to do my part by first writing the chapters; I would then mail the rough draft to Jill and she would read it and note questions and comments. I would then call her to explain and clarify the text, answering her questions. Jill would make appropriate additions and corrections in order to make it read smoothly and clearly, while at the same time keeping my own personal style—if we can call it that! It was surely not easy work, but she set her intention to write it with my tone and as much like me as possible, and she managed to do just that.

Once completed, Jill would fax me the re-written text so that I could make the corrections in my computer, print out the revised writing and send it off to Debbie Darling (Jill's business partner) for editing. Jill introduced Debbie to the SWE tapes and like Jill, Debbie is an avid "fan."

There is no question that these two wonderful beings were the angels I was looking for!

An Invitation to go to Austria

While the book was being created, I continued to speak, travel and work with people on their SoulNote. One of my more exciting invitations to speak came from Luk de Lairess, a man from Belgium who had come to see me when we lived in Bayfield, Colorado. He had his SoulNote done when he visited and was so impressed that he invited me to a symposium on healing that was held in Vienna, Austria in August, 1995.

Naturally I was delighted to go and the trip was fun, challenging, and worthwhile all at the same time. I took copies of a rough draft of my book with me and was thrilled with the positive response it received. A lovely Belgian lady, Florence Fischer, that I met in Vienna, offered to translate my book into French. Also Selina and Uli Steyer, who are from Germany, said they would have it translated into German, and there is a possibility of someone else translating it into Dutch.

Self-Love is the Key

One evening after I had returned home from Austria, I began pondering how, in spite of all the little inconveniences that happened on the trip, it was filled with blessings. It suddenly dawned on me that when I normally travel with heavy luggage, the muscles in my arms are sore for at least a couple of weeks. This time my body felt great after this ten-day trip despite the distance I traveled, all the time changes, and carrying heavy equipment.

There is a reason for this, I believe. For the last ten years, my arms have developed bumps containing some adipose tissue and they look and feel quite bumpy and not at all smooth. Not knowing what to do with them, but realizing that it is nothing dangerous, I have just ignored the bumps on my arms.

After I returned from this trip feeling so well, I was inspired to start acknowledging this part of my body that served me for all these years and is serving me now even better. I realized how I had paid attention to these little imperfections instead of paying attention to the perfection that is there also. It was like paying attention and criticizing the one percent that is not perfect and not acknowledging the 99% that is! As a result, I have learned to accept my body, bumps and all.

Who knows? Maybe the bumps are on my arms to push me to find a frequency that will dissolve them. Like everything else, it is "perfect" if we look at the big picture. Once I focused on the big picture, the perfection and service of my body, I ended up in tears. I accepted and loved my arms just the way they were with their new-found strength *and* bumps.

If we don't love our bodies, they will respond in anger. The main thing is—and it's the key to everything—we must love ourselves and love our bodies—bumps, lumps, imperfections and all!

New Frequencies Discovered

At the beginning of June, 1995, I was given a special formula that would bring the dense physical body into the true Light Body. The tape is a combination of fourteen or more frequencies that need to be played simultaneously. As I explained earlier, normally only three tapes can be played simultaneously in order to get results. They will have a combined total of no more than twelve frequencies at any one time. This new tape breaks this rule a little bit.

The new formula seems to have the purpose of stirring up things and will confuse the body and create a subtle disorder in order to recreate a better order. After I had listened to this new formula of frequencies for a week, I felt something like a little electric shock in the genital area. I am fifty-four years old and went through menopause about three years ago. I never had any discomfort with my period and so I was surprised to experience anything the least bit uncomfortable. But it turned out that the frequencies had triggered the return of my period after only one month of having this electric shock experience. I feel quite certain that my body is reversing the aging process.

I believe our bodies are truly Light bodies, but that before we can become this truth, all parts need to be in harmony, and it's important to do whatever it takes. These new frequencies will help us get there.

I had been given the go-ahead to provide this service. These frequencies are very powerful and need to be closely monitored. These frequencies will only be used in a special location, like at the Awareness Center that I plan to build just as soon as the land manifests. This is the next step, and I look forward to having the center so we will offer people this excellent tool to ultimately manifest the Light body.

This is a very exciting time and what we can do with sound now and in the near future is only a part of this glorious time.

Ascension

Divine Will is to be re-established on earth as it is in Heaven, and we will return to the harmony of being as One.

Since the purpose of Ascension is to spiritualize matter, it is now time to acknowledge and appreciate our physical bodies and everything that is on this planet. The majority of people alive at this time are old souls and chose to participate in this planetary transformation, so it is merely a matter of remembering our true essence. The only thing that is missing to make this earth a complete paradise is more Light. Light removes darkness and helps us to see our true Self. Each moment, Light is showering on the planet and we have the choice to anchor this Light within ourselves and create our heaven on earth, or to let it pass. Whatever we choose is perfect because eventually we will all become ONE again.

The same process of intense heat and pressure that turns graphite into diamond is applicable to our human transformative process. Through our lessons, challenges and growth opportunities, pressure is applied to our bodies so we can raise our consciousness and develop our spiritual abilities at higher and higher levels. Ultimately we each will become the multidimensional Light beings that we really are and **Return to Harmony**.

A Return to Harmony

Part Two

The Recordings

A Celestial Tune Up

Background Information Regarding SWE

It is important to provide some history regarding sound and its uses. Sound is as old as creation. In fact, sound probably *is* creation. Many scholars of ancient languages feel that the causal quotation, "In the beginning was the Word," could have been more accurately translated, "In the beginning was the Vibration," with vibration being the bridge between light and matter. As you probably know, everything is, and has, a vibration. The vibration of Creation may be the original vibration that began everything. Increasing numbers of modern scientists, metaphysicians and mystics now agree that there was vibration before there was matter. So, what does all that mean in everyday life? I believe that it means a great deal!

Some vibrations are more conducive to creation, expansion and evolution than others. These vibrations harmonize and resonate with love, light and life. In the duality of things, some vibrations are more conducive to contraction, destruction, and deterioration. These vibrations are dissonant to love, light and life and would include manifestations of their opposites, such as death, dis-ease, lust for power and control, fear, doubt, guilt and manipulation. The energies of our planet Earth are currently saturated with a vast accumulation of this dissonance and density. The goal of SWE is to help restore the positive frequencies of love, light and life to our planet.

Of course these positive frequencies are still here upon the planet. They haven't left. They are, however, well masked almost everywhere, since the confusions about them are so far-ranging. Now is the time for creating bridges of love, light and life. Massive commitments need to be made to assist all those who want to be assisted, who are seeking to restore the natural order.

As Mother Earth is raising her vibration, we also have to raise ours. We do this by removing all negativity in our lives, at all levels—spiritual, mental, emotional and physical—to "*return to harmony*" and develop the qualities of love, light and life. So I know that working with love, light and life are the most practical issues possible at this time for all of us that want to realize God within.

Realizing That the Power is Within

It is my goal through the use of SWE to help people realize that their power is within. Maybe the time isn't too far off where we won't need tools to help ourselves, we will create everything we want with frequencies and intention. It probably won't be the first time. Many knowledgeable people now believe that some of the great construction feats of ancient history, such as the pyramids, were built with intention and sound frequencies. However, until we are able to use the subtle vibrations of tele-portation and psycho-kinesis, I have developed these recordings as tools for us to use in our own work of personal change and transformation.

What Is Intention?

Webster's Dictionary states: "'Intention' implies little more than what one has in mind to do, or to bring about." This "little more" is the feeling of anticipation or joy that is connected with the thought.

Thought and feeling are both necessary in order for manifestation to happen. Visualization alone will not do it.

The idea of stating an Intention before we play a tape is two-fold. First, we learn to be **specific** when asking for something, by having a clear picture of the goal in advance. Second, we need to realize that our thoughts have created our reality, even if we were not aware of this. Therefore, we need to **control our thoughts**, to really co-create consciously what we want. Someday, when we are able to have only positive intention without any fear or doubts, we will be able to **think** of something, **say** it with **intention** and just **be it**. We have within us all the answers, and now we know the formula:

Frequency + Intention = Manifestation

Listening to the SWE **Chakra (Energy Center) Series** will affect everybody in the surrounding area even if they don't state an intention. The Chakra recordings include vibrations of the body's building blocks—amino acids, hormones, minerals, noble gases and vitamins. The recordings, therefore, are *food for the body*. We cannot avoid being affected by them. Children, plants and animals are living examples of this.

When using the recordings in the **Spiritual**, **Mental**, **Emotional** and **Physical Series**, stating our intention makes a difference. These series will *repair or enhance* the specific bodies. It is helpful to say the intention aloud with each recording, at least to begin with. After awhile, just playing the frequencies can imply the intention without saying it aloud.

It is important, when asking for energy, that a clear intention for *alignment* of the chakras is also stated. This intention, stated in advance, is vital for the energy to be best received and utilized in the body, in a free-flowing, smooth and harmonious manner. Always take a few moments to state your intention for balance and harmony. Then manifestation can really happen.

In the beginning I was using over thirty different recordings with two mineral frequencies on each, and it was simply not convenient or practical for people. The process was limited and cumbersome. I resisted having too many of these recordings and searched for a way to improve and simplify the process. The "Chakra series" was the first step.

Change Beliefs = Miracles

With the physical problems having their origins on the mental, emotional and spiritual levels, I could not see the importance of focusing so much on the physical level. I wondered if by working only on the spiritual level, balance could be achieved. This is the answer I received: "It is a good idea to approach from all sides at once, to provide support for the physical at the same time one is working through the spiritual, mental and emotional levels. It is true that if the spirit itself is in harmony and balance, then all other aspects will take care of themselves. Unfortunately, this is not in our belief system. Most of us at this time do not believe that physical changes can occur simply from belief. Any doubts will prevent what seem to be miracles from happening."

The Standard Self-Improvement Progression

People often ask me what recordings to use for such and such a health problem. I always respond with the same answer: I do not work with problems or disease. I believe the beneficial focus should be on balance and wellness.

This approach is contrary to our usual mode of correcting the problems instead of enhancing our wellness. It's not especially easy to change our approach, but I believe it's necessary.

Think of it this way: Imagine an assessment scale for our physical health of zero to one hundred. One represents being almost dead and one hundred is optimum health. The majority of us are in the fifties on the scale—on the borderline of being sick. We have very little resistance to disease, but we may not yet be in actual pain. Others are below fifty and perhaps are suffering from a chronic disease. When we become stressed, we are plunged down into the forties and experience pain. If we don't make the appropriate corrections, we end up with a chronic disease and then our rating is even lower on the scale. For these people, their primary goal might be to be pain free. This would mean they would only aspire to reach fifty on the scale—a very low goal indeed.

Optimum quality of life occurs when we can maintain a higher level on the scale such as in the eighty to ninety range. At that level, even big stress factors will not make us sick even if we drop ten points down the scale. This means we are able to go through challenges with grace and be more equipped to resolve the related issues.

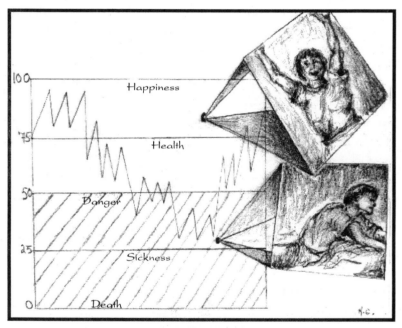

Wellness Scale

The goal of using the SWE recordings is to enable people to improve their well-being so they can maintain a health level in the eighties and nineties on the scale, with the ideal objective being one hundred.

However, wherever we are on the scale today is perfect. If we are in the low range, at one level or another, it serves as a warning indicator. We may need to learn something—perhaps to love and accept ourselves more, to forgive ourselves and others, to change our values, to get rid of fears and doubts, or any number of genuine growth opportunities.

As we get the lessons, we have the opportunity to reach higher and higher levels of relative perfection. We radiate light, love and life and become a living example of harmony, beauty, happiness and longevity.

Ideally, I recommend use of the four Series when a person gets involved with the recordings; however, the use of the **Foundation** (Basic) **Series** of SWE recordings is the *minimum* from which to start.

If your goal is to open your heart, be certain to include the Chakra #4, Relaxation & Calmness, and Love recordings. To release fears and doubts that may be stuck in the solar plexus chakra, include Energy Center (Chakra) #3, together with the Higher Consciousness recording. If you want to make appropriate changes in your belief system, include Chakra #6 with the Brain, Courage & Prosperity recording. In all cases, the remaining Chakra recordings #1, 2, 5 and 7 should be used.

Once the body's energy centers are balanced and energized, and the spiritual, emotional and mental bodies are cleared, then the physical body will have no choice but to return to health and harmony. The Physical Series added to the Foundation Series will, of course, accelerate the repair needed to the specific area in the body that is being challenged.

General Guidelines

I have developed the SWE recordings in order to help people return to and maintain balance and harmony in their physical, mental, emotional and spiritual bodies.

Everything in life, including ourselves, is composed of energy that vibrates at a certain frequency. The higher the vibration, the higher the level. We each have an ideal individual frequency at which we vibrate. Each of our cells and organs has an ideal frequency. Our innate state is that of perfection and perfect or ideal vibrations.

However, our vibratory rates are adversely affected by such things as negative thoughts, dissonant sound, television, microwaves, pollution, stress, trauma, fear, and violence. When our ideal frequency is impacted, disease and negative events can more easily become part of our life experience. The old adage that like attracts like could not be more apt.

To understand how this works, think about a radio. We know that countless radio waves are vibrating through the air all the time. However, we are not aware of these vibrations or waves and cannot pick them up *unless we tune into the specific frequency.* We must match the frequency by dialing or setting it in. Otherwise it does not become part of our reality.

The same is true with our lives. We can only experience those things to which we vibrate or resonate. If we can keep our vibratory rate at a high frequency, then we will not be on the same wave length, so to speak, as an automobile accident, violence, disease or whatever other lower vibratory energy or event you wish to use as an example. Remember, however, that some of the challenges we create in our lives are important tools our soul chooses to experience in order to teach us certain lessons needed for our spiritual growth. There is often an underlying cause for our condition, and the faster we can understand that cause, the better.

The SWE recordings act much like a tuning fork. They remind our cells of their ideal vibratory rate, and the re-entrained cells will begin to match the energy.

Some people have no physical experiences when listening to the recordings, but this is *not* an indication that the frequencies are working incorrectly. Your subtle bodies are actually digesting the frequencies. Just as when you digest food you may not experience physically that it is happening. Nonetheless, you are still converting the food into usable energy for the body. The SWE frequencies work in a similar fashion.

It is strongly recommended that you listen to a combination of the recordings rather than work with just one all the time. The key is "balance," and you won't achieve it with one recording. For instance, a person having difficulty with constipation may be helped with Chakra #1, but eventually, listening only to that recording will put the other chakras out of balance. The frequencies of the precious metals such as Gold, Silver and Platinum, or the noble gases like Krypton and Xenon needed in the higher chakras will not be stimulated, therefore no help will be received for the higher part of yourself.

Just as the physical body requires that you eat a balanced diet and perform a variety of physical exercises, our subtle bodies also need a variety of different frequency patterns to achieve optimum harmonic balance. As you work with SWE recordings, you will develop increased sensitivity to the different frequencies, and discover that at certain times you prefer one frequency to another. You will develop your own taste for the SWE recordings, and your intuition will guide you in knowing when and how to use and combine them.

Healing Crisis

I feel that it is very important to understand the concept of the "Healing Crisis" when beginning to work with SWE recordings. The vast majority of experiences with SWE recordings are pleasurable. Occasionally, however, the frequencies will initially impact some people quite powerfully, causing them to feel uncomfortable, or even sick. The natural response in this situation is to resolve, consciously or unconsciously, not to use that recording again. After all, what kind of person would intentionally do something that causes discomfort?

For an answer to that question, I've asked my friend and expert on the subject, Bob Metteer, PhD, to explain the healing crisis for this chapter.

> *First an answer to your question, Nicole, "What kind of person would intentionally do something that causes discomfort?" For our purposes, it is a person of knowledge, courage, and intention. It is easier to become ever more alive, loving and enlightened when fortified with reliable information.*
>
> *Simply put, a healing crisis is what most of us previously called an illness. In its purest sense, a healing crisis is any condition of discomfort or dis-ease, which means "away from ease." Whether it is caused by stress, trauma, various agents of chemistry, or "bugs," a healing crisis is not fun.*

When we call "feeling sick" an illness, it has an undesirable or negative connotation, something to be avoided, done away with, or suppressed quickly. This goes along with the current western or allopathic medical paradigm which basically says, among other things, "Help the patient to not experience pain." It could be stated as helping the person in crisis escape his or her own feedback. Allopathic treatment assures continued unconsciousness, by ignoring any creative responsibility for sickness, and by suppressing the physical symptoms of sickness with drugs. Holistic health service providers have noted that a health care philosophy which ignores emotion and thought patterns as causal factors just drives symptoms to deeper levels of unconsciousness.

However, if we call "feeling bad" a healing crisis, we have a more optimistic or positive sense that something good is coming of our "downtime." These introspective "blue times," maybe with loss of appetite, achy body, fever, sweating, vomiting, diarrhea, bad breath, body odor, etc. are not symptoms of the body being sick. These are the symptoms of a healthy body detoxifying itself; and such symptoms are more than just the body at work, becoming healthier. A healing crisis can also be an enforced "time out to contemplate" and to "fester" infected and suppressed attitudes of fear and separation up to the conscious level, where they can then be resolved with love, wisdom and power. I offer that a healing crisis is a holistic cleansing process—with body, emotion, mind and spirit acting in concert.

In my personal healing experience, and being a resource for others, I've noticed a monumental step that we can make in self-improvement. That is a shift from seeing ourselves as victims to seeing ourselves as creators, or co-creators, who agree at some level of knowing to go through difficult experiences for instructional purposes. The current, common practice of calling people "victims" of their injury or dis-ease symptoms, reinforces the notion that we're passively at the mercy of the whims of fate. A victim is usually looking for "villains" *upon which to place responsibility for their symptoms.*

By seeing ourselves as creators of our own health emergencies, we can begin to look inward, and see our illnesses as part of a feedback loop meant to help us learn lessons on "Schoolhouse Earth." A common pattern is to be a creator when things are going well but slide into victimhood when we're out of sorts. Unfortunately, the victim/creator choice seems to be an either/or decision. So, when we argue for our victimhood, argue for reasons why we're not the responsible creators of our lives, then we sustain our willingness to be passively reactive and powerless. Instead, we can pro-actively, and creatively, search for self-improvement resources. When we ask our inner guidance for help, then listen and act accordingly, we find ourselves being steered away from conditions of dis-ease and disaster. We're guided instead toward situations of love, healing and appropriate service.

Most experiences with SWE recordings are pleasant and uplifting. However, caution when initially using the recordings is prudent. Some people experience a healing crisis. The **Higher Consciousness** *and* **#21** *recordings need to be used with discretion. Too much too fast has proven to be a physical detoxer for some people. Using* **Relaxation & Calmness**, *the* **Love** *tape, and the* **Chakra Series** *of SWE recordings to start, reduces the likelihood of a healing crisis, as the listener is not in a place of resistance.*

In order to illustrate the potency of the SWE recordings, Nicole has offered the testimonial of Roy Kupsinel, MD, who wrote the foreword to this book. He states:

"Not knowing better, or not heeding Nicole's warning to start slowly with the Christ Consciousness tape, I played it, using auto-reverse, all night, for two consecutive nights. The following day I thought that I had the flu. After one week of the acute phase of a cold and cough, I was left with a zero energy level for the next ten days. I completed my recuperation on a Caribbean cruise. In retrospect, I had experienced a healing crisis—not the flu."

*The addition of SWE recordings to my daily lifestyle has been profound. The frequencies of the Chakra Series are so precise, relating to the minerals, vitamins, amino acids and hormones of life, as well as the vibrations of certain thoughts, so I use them judiciously but frequently each day. Sometimes I play two SWE recordings at a time, or enjoy a music tape with one SWE tape. The #21 tape is one I particularly like, and the **Gateway** tape touches me in a solemn and deep manner.*

I am convinced that these recordings will become a powerful and commonplace growth tool for both do-it-yourself and professionally-resourced health programs. Once the healing crisis concept is truly understood, the listeners will be well prepared. Then when the recordings are played, either alone or mixed with music, they will be very well served.

Happy Listening!

Sets of Recordings

Although there is no wrong sequence in which to play the SWE recordings, there are ways that are generally more beneficial. I have found that the more a person works with them, the smoother the results.

Not surprisingly, balanced use brings everything along harmoniously. Therefore, using the four Series of SWE recordings is ideal. This series of recordings has been available since 1993 and contains recordings that enable a person to raise his frequencies on the physical, emotional, mental and spiritual levels. This, of course, is the primary key to wellness.

For people who cannot invest in all the Series at one time, The Foundation Series of eleven recordings is a good way to begin. It contains the fundamental seven-tape *Chakra Series* as well as *Brain, Courage & Prosperity; Higher Consciousness; Relaxation & Calmness,* and *Love* recordings. This series is a good place to start. It is designed to balance all the body's energy centers (chakras) as well as enhance relaxation, resolve emotional issues, change belief systems, and allow expression of unconditional love and acceptance of yourself and others.

In the Spring of 1995, five recordings were released in order to assist those people who are consciously working with the new energies coming onto the planet, and who want to go a step farther in their personal evolutions. The vibration of some of the recordings was so powerful that even though I had the frequencies, my guidance would not permit me to produce or release them for over two years, because our vibrations as a human race were not high enough.

However, we have now sufficiently raised our vibrations that we can listen to these frequencies and not only be undisturbed by them, but benefit from them. Prior to this, they would have put people into discomfort.

There are a lot of new energies coming on to the planet and although it may not look like it, we have truly made a great deal of progress in clearing out our old personal issues so that we are now in a higher vibration.

In 1997, many other CDs and tapes were created. Now we have four series: the Foundation (Basic) Series, Spiritual Series, Mental and Emotional Series and the Physical Series. Naturally, if you have any physical problems, you should consult your physician along with using the recordings.

How to use the Recordings

1.   SWE recordings **cannot be accurately duplicated**. Without the proper equipment as well as being reproduced in real time, the frequencies will be altered and will not correlate to the healing frequencies.

2.   SWE recordings emit vibratory frequencies that sound much like a hum, a helicopter or purring cat. Each tape is different and emits its own particular sound. Do not think something is wrong with your recordings—this is how they are supposed to sound. There is no music on any of the recordings, but you may play soothing music (please, no rock or other dissonant sounds) simultaneously as you are playing the SWE frequencies.

3.   You may use the SWE recordings while sleeping, doing daily

activities, studying or meditating, etc. You may want to use them at home first before you use them while driving, especially the Relaxation & Calmness recording.

4.   Do **NOT** use **headphones** to listen to the recordings. SWE recordings must be played through speakers in order to receive benefits, since your body needs to experience the vibration of the frequencies. The recordings affect the entire body, not just the ears via hearing. Even a deaf person will receive the frequency waves. The only exception is your SoulNote because you want to learn the special note to be able to hum it. However, you can also play your SoulNote on the sound system.

5.   The recordings may be played at low volume in the background without loss of benefit. There is no correlation between higher volume and increased benefits. The sound does not have to be loud as long as you are around them and your body receives the vibrations. Be sure that some sound comes from your speakers otherwise you will not get any results.

6.   Use of an extended bass or subwoofer on your sound system will enhance the performance of the SWE recordings. Some people use a simple boom-box when they first begin using them, and get results, even if the quality of the unit is not optimal. With a good sound system the recordings are more pleasant, and give faster results. Also, if the speakers produce only the higher frequencies (hertz), the recordings will sound like static and will not be as effective. It is acceptable to begin with any system you have, but a better system is a wise investment. In the future, sound will be one of the major methods used to achieve wellness on all levels, so good sound quality will be essential.

7.   Do **NOT** play the SWE recordings while watching television. TV waves negate the benefits of the frequencies and playing them at the same time is futile. However, you can be in the same room where the television is playing as long as you are NOT watching it.

8.   The SWE recordings will benefit any person or pet living in the

home. The ideal circumstance is to be in the same room that the recordings are playing. However, it is workable to play the recordings and still get benefit if the frequencies have to travel through a single wall.

9. SWE recordings are ideally played on an auto-reverse tape player *except for the Higher Consciousness tape and tape #21.* (See The Emotional Series in Chapter Eleven.)

10. You may play the recordings one at a time or in sequence, e.g. Chakra #4 and then the Love tape. Or you may prefer to play them simultaneously if you have the equipment. Both ways are effective. However, playing the recordings simultaneously might bring results more quickly. **Please do not play more than three recordings at a time,** since more than three confuses the body.

11. A good way to work with an area with some challenges is to play the Relaxation & Calmness recording first. It is important that you play the other recordings also, especially for the chakras, so you don't concentrate on the problem and you stay in balance.

12. Play every Chakra recording *in its entirety* before using the next one, because each sequence is unique. A sequence may contain up to four different frequencies. For instance, Chakra #7 has twenty-four sequences containing eighty-seven frequencies. For the other series, the recordings are also composed of different frequencies, but because they repeat themselves they can be listened to one side at a time. You may want to do this at the beginning if you feel listening to the whole tape is too strong. When you resume listening, don't rewind, but continue from where you stopped. Listening to both sides is the ideal.

13. When using any of the SWE recordings, it is recommended that you *say out loud the INTENTION* related to each one (see Appendix C). By doing so, you will maximize results as well as participate in the co-creation of your new states of wellness. You can also

change the written intention and create your own if you wish. This can only be more powerful because it will be precisely adapted to your needs. What counts is YOUR intention.

14. Specific recordings work well together and create a synergism. By combining a Chakra recording with a Mental, Emotional, Physical or Spiritual recording, the processing and healing will be more gentle and effective.

Suggested Combinations

On the following pages you will find the generally recommended combinations in which to play the recordings. However, the SWE frequencies can be listened to in any order, and the combinations which follow are a guideline only. Please know that your inner guidance may lead you to other combinations. I encourage you to tune into what's best for you.

Recordings and Recommended Use

The Foundation Series:

Contains 11 Tapes or 12 CDs: *Recommended Combinations:*

Energy Center (Chakra) Series   Chakra 1 and 7
  plus   Chakra 2 and 5

Brain, Courage & Prosperity,   Chakra 3 and Higher Consciousness*

Higher Consciousness,   Chakra 4 and Love

Relaxation & Calmness,   Chakra 6 and Brain, Courage &

Love   Prosperity

                                   Relaxation & Calmness**

The Foundation Series with the Spiritual Series:

Contains 12 CDs/Tapes: *Recommended Combinations:*

Thymus   Chakra 1 and Patience

Transition   Chakra 2 and Relaxation &

MerKaBa   Calmness**

Energy, DNA, Enzymes   Chakra 3, Higher Consciousness*,

Patience   and Energy, DNA, Enzymes

Oneness   Chakra 4, Love, Connection,

Purification   and Shifting Consciousness

Gateway   Chakra 5, Thymus, and Purification

Immortality   Chakra 6, Immortality,

Connection   and Brain, Courage & Prosperity

Shifting Consciousness   Chakra 7, Transition, Merkaba,

Manifestation   Oneness, Gateway,

                               and Manifestation

---

*   Higher Consciousness should be played just once through each day.

**   Relaxation & Calmness may be used alone or with any CD/tape where resistance or discomfort is felt or when your intuition calls for it.

*** Cell Rejuvenation can be used with any energy center (Chakra) needing help.

**** Prana can be played endlessly as it helps us breathe. This is especially helpful at high elevations.

~ The recording entitled #21 should be played for just a few minutes at the start, then slowly increase the time. The frequencies on this recording will bring mental, physical, emotional and spiritual issues to the surface, and you will want to confront them and resolve them harmoniously. . . and not too fast. Your comfort level and intuition will be your best guide.

The Foundation Series with the Mental/Emotional Series

Contains 12 CDs/Tapes:
SpaceLight
Safe Environment
Clarity/Focus
Communication
Choice
Memory
Public Speaking
Body Wisdom
Addiction
Emotional Release
Brain/Body
#21-Light Transformation*~*

Recommended Combinations:
Chakra 1 and 7
Chakra 2 and Relaxation &
    Calmness
Chakra 3, Higher Consciousness,*
    and #21*~*
Addiction, Emotional Release,
    and Memory
Chakra 4 and Love
Chakra 5, Communication,
    and Public Speaking
Chakra 6 and
    Brain, Courage & Prosperity
Clarity/Focus, Choice, Body
    Wisdom, and Brain/Body

Safe Environment and SpaceLight do not need to be accompanied by a Chakra Recording. They may be played alone or with any combination.

The Foundation Series with the Physical Series:

Contains 12 CDs/Tapes:
Prana
Hair
Circulation
Cell Rejuvenation
Bones
Hearing
Hydration
Compassionate Heart
Pancreas
Perfect Veins
Stability & Stillness
Vitality & New Health

Recommended Combinations:
Chakra 1, Bones, and Hair
Chakra 2, Cell Rejuvenation,***
    Pancreas, and Vitality & New
    Health
Chakra 3 and Higher
    Consciousness*
Chakra 4, Love, Prana,****
    Compassionate Heart,
    and Stability & Stillness
Chakra 5, Circulation, Hydration,
    Perfect Veins, and Hearing
Chakra 6 and
    Brain, Courage & Prosperity
Chakra 7 and
    Relaxation & Calmness**

The four Series contains 47 tapes or 48 CDs, in the following combinations:

Chakra 1, Hair, Bones, and Patience

Chakra 2, Cell Rejuvenation, Pancreas, and Vitality & New Health

Chakra 3; Higher Consciousness; Memory; Addiction; Energy, DNA, Enzymes; Emotional Release; and #21

Chakra 4, Relaxation & Calmness, Love, Connection, Shifting Consciousness,Compassionate Heart, and Stability & Stillness

Chakra 5, Prana, Circulation, Purification, Communication, Public Speaking, Thymus, Hydration, and Perfect Veins

Chakra 6, Brain, Courage & Prosperity, Clarity/Focus, Immortality, Hearing, Choice, Body Wisdom, and Brain/Body

Chakra 7, Transition, MerKaBa, Oneness, Gateway, and Manifestation

The Recordings

Details About Each of the Sound Wave Energy Tapes and CDs

For six years, we have been recording the SWE frequencies on audiotapes, because the sound that comes from an audiotape contains harmonics and has certain qualities that make it more organic. I have used the best quality metal tape to prevent stretching, and if by accident a tape breaks, the client can return it for replacement. People continue to receive numerous benefits from the SWE tapes, and it is my intention to continue with this form of production.

Audiotapes Do Have Certain Drawbacks

1. The tape deck is not always calibrated perfectly, and needs to be cleaned regularly.

2. The companies that produce tape decks produce only those that can play two tapes, one after the other. Since we have many tapes to work with, they must be changed often.

3. Sometimes the audiotapes are too tight and need to be fast forwarded or rewound.

4. Tapes can be erased if placed near a magnet. Also, they sometimes have to be demagnetized. At the same time, audiotapes need the earth's magnetic field to retain their data.

The above conditions make audiotapes somewhat fragile. However, from my research, compact disks (CDs), as they are normally produced, would have provided only around 77% of the results of the audiotapes. Because I did not want to compromise the results, I dealt with the various drawbacks of the tapes. Also, whenever I asked my guidance if it would be better to use a CD instead of an audiotape, I received a "no." I thought CDs were not an option, until the technology for creating CDs changed.

One day, I asked if there would be a way to produce CDs that would have the proven results of the audiotapes. The answer was "yes." To do so, we had to use the most current technology available, and add certain components that would make CDs as effective as the audiotapes. The technology that creates a CD is based on a digital signal, and it does give a very clear signal, but it doesn't produce the harmonics as does an audiotape. This is why many music lovers prefer audiotapes to CDs. The majority of people, however, believe that a CD is "better" because the sound is very clear. But clarity is only one aspect of the whole picture.

One of the components we had to add was to put in subliminal messages at the same time as the frequencies. I had been afraid of subliminals (inaudible words spoken on a recording, underlying the predominant sound) since I was first aware that they existed, which was around thirty years ago, so I did not feel comfortable with the idea. I knew their power and how some have used them to control others in their decision-making. I was not interested in having ideas put in my brain without my consent. I could not think of any positive side this tool may have, so I resisted this technology.

Since I had asked my guidance to give me a solution to the problem of CD vs. audiotape, I had to confront this fear and realize that, like any other technology, subliminal technology can be used to control or empower people. So, I agreed to place the "Intentions" of the frequencies on the CDs, in a subliminal manner. Because of current CD technology, we had no choice. However, anyone who objects to subliminals can still use audiotapes.

So, now I have cleared all of my objections to subliminals.

To produce the best CDs, specific colors and special geometric shapes were added. The company that produced the CDs didn't understand why I wanted colors that were not visible. They were not accustomed to placing three colors on top of one another, with only the last one visible. But they learned it was all part of the process. I did everything to make the CDs as powerful as the audio-tapes, and I am very excited about them! I learned from Chuck Pleasance, a leader in "mastering" Cds, how to do it. I personally did all of the subliminals and was present in the mastering process (creating the master CD). I even did a portion of the mastering myself!!!

The Advantages of CDs

1. You can purchase a changer that plays all 48 CDs (without you doing a thing).

2. Because CD players are very popular, you have a bigger selection and will be more apt to find the equipment that best corresponds to your needs.

3. CDs are not affected by magnets, so they don't have to be demagnetized.

4. They cannot be accidentally erased.

5. The CDs come in albums of twelve and take up half of the space of the tapes.

6. If you don't say the "Intentions" when you begin to listen, your subconscious mind will receive it anyway (in French and English). I say all of the "Intentions" with all of my heart, in each CD.

In 1998, with this revision of "Return to Harmony," we introduce the SWE CDs. It is another step towards bringing this technology to more people around the world.

SWE produces 47 tapes or 48 CDs. Because a CD can record a maximum of only 74 minutes, the Chakra # 3 is in two parts on CD. The CDs and tapes are divided into four different series. Each series will assist you in achieving specific goals:

- Enabling the body to assimilate all building blocks needed to nourish and balance the endocrine system (glands). Let go of resistance, old belief systems and stuck emotions, open to unconditional love of the self with the **Foundation Series.**

- Accept miracles, connect with angels and your Higher Self, enhance your healing abilities, feel oneness, release entities, integrate new energies, build a strong immune system, promote awareness of the immortality of your being, give energy, create light around the body, promote patience, balance ego, attune to the greater purpose of humanity, support love in action and assist in manifestation with the **Spiritual Series**.

- Release fear, doubts and anger, bring Light into your bodies and into your space, enhance public speaking abilities, encourage body wisdom, enhance communication, release addictions, focus on your mission in life, experience clarity, balance the right and left brain, improve memory, and BE in the moment with the **Mental / Emotional Series**.

- Balance brain chemistry, circulation, bones, hearing, heart, pancreas, veins, small intestines; promote hydration, good breathing, acid/alkaline balance; repair and rejuvenate the body with the **Physical Series**.

THE FOUNDATION SERIES

The Foundation Series (formerly named the BASIC SERIES) is the groundwork for the SWE technology. Each person interested in this system works with the Foundation Series combined with the other series. The **Foundation Series** contains eleven audiotapes or twelve CDs, including the seven Chakra (Energy Center) recordings, Relaxation & Calmness, Love, Brain, Courage & Prosperity and Higher Consciousness. (The CD series has two CDs for Chakra #3.)

The energy centers of the body, or chakras, simultaneously transmit and receive Universal Intelligence, so it is critical that they be open and balanced in order to co-operate with the basic elements of love, light, and life. Our energy centers are severely impacted by the negativity in our lives, so it is not unusual for them to become unbalanced.

Energy Center (Chakra) #1 contains 42 different frequencies, and **Energy Center (Chakra) #7** contains 87. I included the necessary frequencies of vitamins, minerals, amino acids, noble gases and hormones related to each center in order to provide all the building blocks necessary for their optimum performance. When the centers have everything they need, they can resonate properly and tune up the entire system.

As each energy center begins to resonate with its own healthy frequencies, it will naturally begin to align with all of the other centers. As they come into alignment, our physical energy, feeling of well-being, and inner strength will be enhanced. When the energy centers are in balance, these effects are a natural result, and we are in harmony with our True Self as well as with our earthly environment.

The best way to easily strengthen an area is to use a specific energy center (chakra) recording which corresponds to the area of the body that needs support and change. For instance, if you want to improve your hearing, you would benefit from listening to the **Energy Center (Chakra) #6** recording, then **Hearing,** followed by **Cell Rejuvenation**. These recordings, like any of the SWE recordings, may be played in a single sequence or can be played simultaneously. Please do not play more than three tapes or CDs at a time, since more than three may confuse the body.

ENERGY CENTER #1
Base or Root Chakra
Grounding

The **Energy Center (Chakra) #1** recording represents our connection with and grounding to the earth. It also contains the Kundalini, or life force, energy that sustains our physical bodies. This first center, or chakra, with its grounding aspect, is the seat of the Will to Live. It also governs the eliminative functions of the body.

When this chakra is out of alignment or out of balance, we experience feelings of insecurity and vague longings which cannot be clearly identified or satisfied. Imbalance in this area leaves us with a confusion regarding our place in the physical environment.

Once this center is realigned and it is resonating with its ideal natural frequencies, not only will our physical bodies be enhanced, but we will feel secure and safe and fully in creative control of each situation that arises in our physical environment.

These are the frequencies to ground oneself. When combined with the *Bones* recording, the legs, pelvic region and spine benefit. These regions basically comprise our support systems.

Many people have concentrated on developing the higher chakras, (four, five, six and seven), thinking they are more important. The truth is, all of the chakras are equally important, and it is disempowering to focus on one or one group of chakras at the expense of the others. Not surprisingly, the key is to balance all of the energy centers so they are all open, balanced, and in harmony. Sometimes people erroneously believe that because the first chakra is associated with money, the second with sex, and the third with power, they should be ignored. Just because there have been instances of abuse of money, sex and power on this planet, it does not mean we should reject them altogether.

It's important for us to remember that our soul chose to experience planet Earth in a physical body at this time. As Tielhard de Chardin said, ". . .we are spiritual beings having a human experience," not the reverse, as is so commonly assumed. As spiritual beings, we have chosen to inhabit a human body, and all aspects of being in a physical body are inherent in creating balance. It is wise for us to take care of our physical body, otherwise we will be unable to achieve our purpose for incarnating here. We are here to spiritualize the body, not to ignore it.

ENERGY CENTER #2
Sacral and Spleen Chakra
Balancing Masculine and Feminine

The **Energy Center (Chakra) #2** recording energizes the sacral and spleen areas of the body. This second chakra governs the reproductive organs, ovaries, uterus, and testes and regulates the supply of oxygen to our body. In addition, it regulates the supply of new blood cells and the life force energy to the body, and is the seat of "polarity." When aligned, it regulates the energy of gender balance, the masculine and feminine aspects within each of us. When this energy center is balanced, a loving relationship is created with our self and others.

The Chakra #2 recording contains the frequencies of hormones such as estrogen and progesterone as well as the minerals, noble gases and vitamins found most abundantly in those areas. My guidance told me that the testosterone level on the earth is already too high and as a result, this hormone is not included on the tape.

When the second chakra is in balance, it especially benefits the kidneys, pancreas and other organs related to that area of the body. This center needs to be working properly for the rest of the body to be in harmony and balance.

I feel it is especially important for the masculine and feminine energies to be in balance at this time. The planet is now emerging from a period where masculine energy was the dominant influence. This gave us an excess of will and purpose, or *yang* energy. Yang energy dominated at the expense of *yin*, or feminine, energy, which nurtures and protects. A restoration of balance is needed.

At this particular point in our evolution, it is extremely important for souls who currently inhabit male and female bodies to recognize that we each have both feminine and masculine energies within ourselves. We must be able to access both of these qualities in order to achieve balance, understanding, and wisdom.

Such awareness and balance will allow women and men to work together in ways that have not previously been possible. With this shift, things will be much easier between the two sexes and strife and misunderstanding will be dramatically reduced. Each of us will recognize within the opposite sex a strong sense of self and oneness. The previous illusion of seeing ourselves as separate will dissolve and we'll enjoy a new, profound understanding. This is the most important aspect of this second chakra or energy center.

This second chakra is one of the major problem areas for women. This is not surprising when you consider how male-oriented and dominated our society has been. Women often have not felt comfortable being women in a society where masculine qualities were prized and feminine ones were barely accepted, if at all. What happens when a part of our identity and our body is rejected? The body responds to rejection or lack of acceptance with dis-ease. This may manifest as difficulties with menstruation as well as problems in the uterus, cervix, and breasts. Many other areas can be negatively impacted as well. The more we ignore our body or, heaven forbid, hate our body, the more it "cries out" for attention.

The primary need of any body or body part or organ is love. Unfortunately, love is often the very last thing we give our bodies. The solution is acceptance and love. Once we shift into the love mode, then we can give our bodies some physical help. But change for the better is simply not possible when hate, resistance, and fear are in the picture. Thus, the critical first step toward change is the unconditional acceptance of self.

Think of your body or a part of your body that has been ignored or rejected as if it were a little outcast child, one that is not wanted by the family. When such a child is not loved for who he or she is, the child will attempt to get attention and love by doing nice things at first, but if this does not work, the child will create all kinds of difficulties to get some attention. The errant child becomes a pain. The manifestation in your body is identical. The unloved body becomes a pain and expresses dis-ease.

If you have been heaping scorn and ridicule upon your body,

or even just ignoring it, then I advise going through a process each day by which you literally give yourself a mini full-body massage and thank each part as you are massaging it.

For example, you might start by rubbing the soles of your feet and consciously thank your feet for supporting you today and taking you where you wanted to go. Then move to your ankles and briskly rub them as you thank them for being the critical link between your feet and your legs. Go over each body part this way and thank each part for the service it has provided you that day. You might thank your arms for enabling you to hug your children today. Be certain to include your head and face in this loving rubdown. For an extra-added benefit, pause and thank your ears, eyes, nose and mouth for service rendered, as well as the major organs such as your heart, lungs, liver, intestines, kidneys, pancreas and spleen. Thank each with sincerity and love and your health is bound to be better and better.

ENERGY CENTER #3
Solar Plexus Chakra
Seat of Emotions and Personal Power

The **Energy Center (Chakra) #3** recording corresponds to the solar plexus. The solar plexus center is the seat of our emotions and personal power. This is the area that relates to the source of our "being." Sacred Geometry tells us that at conception our original eight cells were located in this area, at the level of the navel. I was taught that these eight cells don't ever die, they just transmute or change to another form but stay there nonetheless.

I don't have absolute proof of this, but I can tell you that when people have experienced the *Oneness* recording, over half of them have told me that they experienced oneness in their body around the belly button, or solar plexus, area. This is where we feel connected. It's like the brain of our "gut feeling."

Emotions are extremely powerful and are one of the two essential ingredients for manifestation. The finest visualization that a person can conjure is worth nothing if it lacks the power of feelings or emotions. Thoughts and feelings are living vibrations, therefore it is very important to learn to manage or control them so we can *be in power* in our life. The fact is, we have the choice to use the power of our emotions to destroy and bring more separation, or to use it consciously to manifest a wonderful life.

The only way we can restore real personal power—**the power of love** in our life—is to cleanse this solar plexus chakra of all our negative emotions of fear, doubt, guilt, envy, etc. Remember, fear is an emotion and a feeling, and if this is our focus we will manifest fear-based experiences in our life.

The purpose of the body's third energy center has been so misunderstood and abused that the majority of people are completely out of balance in this area. Most have simply given their power away, just handed it right over. A small percentage of people in this world make it their business to take control and attempt to manipulate and disempower the majority. This small but significant group of people is at one extreme end of the spectrum. A much more common pattern is for people to give their power away to one or more institutions or systems such as governments, churches or schools, to name a few. Those people and institutions that wish to usurp power exercise their control using fear, threats, drugs, brainwashing and violence. Our emotions become confused, and in this confusion, we are prevented from remaining connected to our "gut feelings." Gut feelings help us remain connected to our seat of personal power. When we are separated from our gut feelings, we have lost personal power and are at great risk.

Karl Marx, the father of communism, said that "religion is the opiate of the people." But that statement doesn't apply only to some organized religions; more significantly, it applies to prescription drugs as well as street drugs. Everything and anything that is used to numb us and prevent us from raising our consciousness falls into the same category. Such things as alcohol, food, work, and sex serve to put us into altered states and lower our vibratory rates, taking us farther and farther away from the recognition of our connection to the Divine and to each other.

Each time we are attached to anything (particularly an addiction or an obsession), we give it power and siphon off our own personal power. The way to take our power back and keep it is to *become detached.* Even being too busy to think, because we are working hard to earn money for survival, is an attachment that dilutes our power. It's disheartening to realize that a number of things prevent us from taking the time to become truly aware of how we have lost personal power. The result is that we remain lost.

Taking responsibility for our lives means that we do not blame others for what happens to us, no matter how difficult the situation. Worst-case scenarios like abuse are no exception. Personal responsibility is still the optimum. I believe each soul chose what it needed to experience for its greatest learning, and this includes abuse, as hard as it may be to understand. The moment we decide to take responsibility for whatever situations we have created in our lives, we retrieve our power. Think of it from this standpoint: has a single human being ever felt a surge of power when they have been called a victim? Quite the opposite. When we slap on the victim label, we instantly feel weakened and diminished. The way to empowerment is to take responsibility and cast aside the victim label. The magic begins when we take responsibility. Sometimes it is frightening, but it is the only way to really make essential changes. When we expect others to change instead of ourselves, we are courting disaster. We can only change ourselves. Each person is responsible for him or herself, it can be no other way.

Can a Belief Create a Disease?

For over forty years, I regularly had what is called a "cold." I would get the cold once or twice a year and it would last three to four days. One day I read that what we call a cold is really a cleansing of the mucous accumulated in the many canals connecting the nose, throat, ears and sinuses. After believing and accepting this theory, I no longer had colds. I reduced my "cold" from three or four days to a "cleansing" that lasted only two hours. This came about because of my power of belief. I really believed that it was a cleansing and not a cold. Since that time, I've never had another cold. Instead, maybe once a year I will have a couple of hours where my nose will run, but that is all. I took responsibility for what I believed and what my body manifested. This made it possible for me to change the pattern and no longer be the "victim" of colds.

During the many years that I worked at the hospital, many gynecologists prescribed annual mammograms in order to detect cancer. The first year that a woman would come for the mammogram, her results would show nothing. However, just the mere fact that there was a formal "search" for disease set up a negative expectation. All too often, subsequent visits manifested exactly what was being looked for: disease, not wellness. This routine, although based on a sincerely well-intentioned premise, was a good way to ensure that something developed. We know the debilitating effects of radiation, and when you combine that with a fear of cancer triggered by the x-ray, you have a recipe for disaster. All the ingredients are present to create cancer in the years to come, and, in many instances, this is exactly what occurs.

I certainly do not encourage the denial of a condition. I believe there are other ways to be aware of our body's lumps and bumps: touching is one of them, and it is simple and much easier.

Pain allows us to become aware that there is an imbalance in our body. It is important to acknowledge imbalances, but not to dwell on them. Remember, we get more of what we focus on. When we give our power or attention to the imbalance, we commit ourselves to remaining that way. We are constantly reaffirming the diseased condition by focusing on the disease. By labeling the problem, we have a tendency to identify with it. Another good example of this is the name that we give to people who have contracted certain dis-eases. Each time a person states, "I am asthmatic," the unbalanced state is reaffirmed. It would be more to our benefit to dwell on the positive changes that will bring balance and a return to harmony.

Whenever we identify with something we give it power. "Where the attention goes, the energy flows." This emotional link may be a profession, a partner, a situation, a problem, or a disease. "I am a truck driver," "I am bankrupt," "I am homeless." When we make these kinds of identifications, we often lose our sense of self and therefore our personal power. Any of these links become our truths. They limit our "beingness." When we become detached rather than attached, we learn to manage our emotions. The more detached we are, the more *powerful* we become.

What Programming Do We Favor?

Another important aspect of this whole concept is that we refuse to focus on any of our weaknesses, because what we focus on, expands. For example, it has been proven that if we constantly repeat to someone that they are stupid, eventually they will believe it and act accordingly. If we say destructive things to ourselves time after time, then the thing we tell ourselves comes into being. We *become* it. This is called negative programming. Even if it is a joke—the subconscious mind does not know the difference between a joke and the truth.

There is only one power and that is the power of love. All "positive" thoughts of harmony, balance, beauty, joy, happiness, abundance etc. are our birthright—and what we can choose to bring into our lives. Positive affirmations *with feeling,* pave the way to a joyful, healthy life.

So many of the problems of the human condition are seated in the solar plexus chakra. The causal factors related to our misery, unhappiness, and fears are stuck in this energy center. It is critical to release the negative emotions related to this center. Playing the Energy Center (Chakra) #3 recording gives us the strength needed to work with the *Higher Consciousness* recording, which will bring all emotions to the surface that need to be cleared.

The solar plexus center also governs our digestion. So if our emotions are suppressed, confused, fearful and repressed, there's a good chance we will have digestive difficulties such as stomach aches, diarrhea, constipation or even more serious ailments. When we are balanced and in power, we can digest, absorb and assimilate not only food, but the gifts and challenges from life as well.

This chakra is particularly important at this time of change. The entire physical body can be pushed out of balance and out of focus by external events. It is important for the physical body to be nourished because when the physical body is unwell, it is difficult to receive the gifts from Spirit. So the well-being of the physical body, which is provided by this third chakra, is extremely important. It's very difficult to concentrate on feeling good when we are in pain. Eventually, all who want to will learn to listen more closely to their inner guidance to determine what to eat or drink, and what to skip.

This energy center is also the seat of our feelings. It is the center through which most mediums work and clairvoyants function. Our "gut feeling" is one of the four ways that our soul can communicate with our physical self (along with clairaudience, clairvoyance, and intuition). Therefore, each of us can achieve balance and resonance from our inner guidance when this center is balanced.

There is a profound benefit to feeling well, and this third chakra provides the support base for feeling happy, which occurs when it is in true balance. The energies of the solar plexus chakra have to be directed to the heart center where the power of love abides. The desires of the little self will then be transmuted into service to the Higher Self.

By the way, the Energy Center (Chakra) #3 recording is the only one that is 90 minutes long, the rest are only 60 minutes. This protracted length is essential. Guess why!

ENERGY CENTER #4
Heart and Thymus Chakra
Where Love and Courage Abide

The **Energy Center (Chakra) #4** recording corresponds to the heart and the thymus. It also governs the blood, circulation, lymphatic system and the lungs. The lungs cleanse and purify; the heart expresses love, wisdom and courage.

The thymus is the endocrine gland which acts as the master computer for the immune system. With all the viral diseases that are around, a powerful immune system is necessary. When the thymus is developed, it becomes the seat of **unconditional love**.

As we raise our frequencies or vibrations, we become Beings of Light and manifest our true selves. The heart/thymus chakra will then be the physical center of our being. It will connect the heart and the throat chakras and help develop the loving communication between the lower self, the Higher Self and our angels and guides.

The next connection we make as we raise our frequencies will be between the third, or power, chakra and the sixth or third eye chakra. This enhanced cooperation will help develop our telepathic powers. This will be followed by the raising of the Kundalini by uniting the first, second and seventh energy centers to help us reach enlightenment. We will then be fully aware that we are Beings of Light.

The Chakra #4 recording contains a specific frequency to accelerate the opening of the thymus. This frequency plays for five minutes on this tape. The process I went through as I developed these frequencies was quite interesting. I knew there was a specific frequency for the thymus. I was working with numbers to find the frequency, and my guidance told me it was between 200 and 300 hertz. This was a big help, but nonetheless, it took me over a month to find it. As you may know, when we receive guidance, whether it comes from ascended masters, angels, our personal guides, beings of light or whomever is of the Light, we are always empowered so we don't depend on them. If they were to tell us everything, we would become dependent and never trust ourselves. Hence, we'd never come to know that deep place inside ourselves that we already know but have just forgotten.

The mere fact that we think of something means we have the ability to achieve it. We all have this ability, there is nothing magical about it. So trusting and knowing that we have all that we need inside us is surely a good place to start. When we are quiet and pay attention to our inner messages, we will eventually connect to our heart's desire. It is the key to whom and what we are.

I originally intended to place the growth hormone frequency with the Chakra #6 recording, where the pituitary gland is located, because this is where the growth hormone is produced. But I was guided to include it in the Chakra #4 recording instead. My guidance explained that the collective belief in our culture is that we will naturally die when we are seventy to eighty years old. This belief means that the "growth" hormone essentially deteriorates into becoming the "death" hormone, which encourages decay and demise. However, if the frequency is placed in the heart center (the fourth chakra), the seat of the soul, which knows we never really die, then the growth hormone will continue to be produced and will encourage life and growth.

The growth hormone not only physically rejuvenates our cells, but also provides necessary energies for the courage to grow with Spirit. It is not only a physical hormone but a spiritual hormone as well. Its connection to love and the heart center is vital. Without it, growth is not possible in the way that is ideally intended.

The Energy Center (Chakra) #4 recording has beneficial effects on our circulation, lungs, and use of oxygen. It also strengthens the circulatory and respiratory systems so that unhealthy particles that are breathed in can be easily eliminated. The skin will rejuvenate as the cells better utilize the nutrients provided. Unhealthy toxins are eliminated in cooperation with organs energized by the *Energy Center (Chakra) #1* recording. Increased circulation and the improved use of beneficial elements have a very positive effect upon the body's appearance as well as on our skin and hair.

It is OK to Cry When We Are Sad

In our society, the majority of men believe that crying is a weakness. When a little boy hurts himself or is separated from a parent, friend or pet, and responds with tears, he is quickly admonished with the terse reminder that, "Men don't cry... now be a man and stop crying."

As a result the little boy, who wants to become a man, tries to suppress his feelings and learns not to cry. He surely doesn't want to disappoint his parents, so he adopts a stoic response to emotional triggers. The child grows up holding high the goal of being "tough." Because he denies the tender aspects of his being, when he begins to be interested in the opposite sex, he focuses on performance instead of the intimate involvement required to create a deep, meaningful relationship.

Little boys that have an inclination to play with dolls are told that boys don't do that—with the strong implication that it is wrong. So the boy feels guilty and will express the emotion of anger with one of his toys, and this will be seen as acceptable. Of course, many of the toys the little boy is given encourage him to "play" at killing, destroying—and surely not loving.

Few parents realize how critical such early influences are on a child's development, and if they do begin to realize it, the damage is already done. It's a repetitive cycle for many parents who simply do to their children what was done to them. No one is really to blame. This is simply part of the mass consciousness, and we all are affected by it. We are at the mercy of this energy if we're not aware that it exists; but once we're aware of its existence, then we can make a choice, forgive ourselves and change our own beliefs. There's no one to blame. The only course is to take responsibility for ourselves and our own beliefs. Once we make the change, people around us will also make changes based on our example, just as a ripple in a pond continually widens and expands its presence.

The Energy Center (Chakra) #4 recording is probably the one that brings up the most resistance. While it opens the heart, it also requires us to alter our beliefs regarding the aging process and the denial and suppression of feelings. We need to have an open heart in order to live longer and be happier. The Chakra #4 recording can help this process.

ENERGY CENTER #5
Throat Chakra
Speaking Our Truth

Energy Center (Chakra) #5 recording corresponds to the thyroid and the parathyroid glands, in the area of the throat. It controls the vocal area, respiration and metabolism. It guides peace and trust of the heart into creative expression. This center is the outlet for both the positive and negative within and helps us to express both—which is our truth. It also regulates the metabolism.

Many food substances are ingested during the course of the day. Some of these have great value, but many do not. It is important that each organ of the body be in balance so that each can recognize needed nutrients when they are available. This fifth chakra reminds the thyroid of the frequencies of balance and ensures that it does what is necessary to restore harmony. The thyroid gland controls the rate at which nutrients are metabolized and made useful to the body. The effect of the food energies that enter our bodies as a result of eating is therefore extremely important for the thyroid to balance.

The thyroid center (or throat chakra) also assists us in speaking. All things that are spoken as one's Truth resonate with one's self. When things are in harmony with Self, the feeling of openness in this center results, for there is nothing to hide. There is never stress or a feeling of soreness. The frequencies used in this recording enable one to feel balance and harmony and also to increase clarity of speech.

While men were told it was not okay to cry as children, they were allowed to display anger, rage and even fury and violence. Men have used this way of expression not because they are bad, but because they were taught to behave this way by their parents and society in general. If they didn't want to be vulnerable and continually hurt, they had to shut down their hearts. Even if someone beat them up, the accepted way for them to show their pain was through anger. Everything was externalized.

For women, it has been the opposite. Women have been allowed to cry when they were little girls and as grown-ups. Expression of feelings and emotional reactions were encouraged and were internalized. It has been permitted and idealized that women focus internally and feel what's really inside. However, women were most decidedly not permitted to express other emotions like anger, indignation and rage. They were seldom encouraged to speak their total Truth when it involved "male" feelings. Women have been conditioned to speak only "niceties" and not to demonstrate the so-called negative emotions. Therefore, words not spoken and emotions not expressed but repressed stayed inside and eventually block the throat center, making it impossible for a woman to speak her Truth.

Because women keep their hearts open, they know how it

feels when somebody tells them something that is hurtful. Being afraid of hurting another, they often speak only part of their Truth, not fully saying what they really think and feel. In this way, women become hypocritical without ever intending or realizing it.

Many of us are now becoming aware that we want to stop conforming to the model society has imposed on us, and be who we really are. All of us, men and women alike, will benefit from acknowledging both the feminine and the masculine aspects in each of us and totally accept both. As we open our hearts and let everything in our lives be guided by our heart and love, we will be able to say everything we think. If we truly come from the heart and from love, whatever we say will be ideal. At first this may be difficult, and because women have little practice in speaking their Truth, words may come out in a less than loving way. But we will learn, and it will get easier and better. Men, on the other hand, may experience some pain while they are opening their hearts and keeping them open. But, the joy will be greater than any pain, and therefore worth going through the process.

ENERGY CENTER #6
Pituitary / Brow / Third Eye Chakra
Knowledge

Cosmic life energy enters the body through the sixth chakra and energizes the pituitary gland and the third eye. This energy center also governs the eyes, ears, nose, teeth, nervous system and lower brain. The brow, or third eye chakra is the seat of All Knowing. It also regulates acidity and alkalinity levels in the body.

When the pituitary gland is in balance, it appropriately secretes an ever-increasing number of identified hormones. These hormones chemically "talk to" the other endocrine glands. When this energy center returns to harmony, the third eye also opens, enhancing clairvoyance.

Today more people than ever are experiencing an opening of their third eye due to our increased vibratory rates. This higher frequency allows us to tune in to the finer energies. Many people are focusing on increasing their clairvoyant abilities; some people had this ability in past lives and are awakening to these old abilities. It would be wise to be aware that before long all of our thoughts will be public domain as telepathic powers become more developed in many people.

The frequencies used for this brow center, or sixth chakra, help create a doorway to our higher understanding and knowledge. This creates a window to the knowledge of one's self, often including information about past incarnations. When attention is focused, using this tape, wise use of knowledge can be accessed, understood, felt and seen as the brow center is harmonized and balanced.

ENERGY CENTER #7
Crown and Pineal Chakra
Union of Body and Spirit

The seventh chakra corresponds to the crown energy center. It also balances the eighth and ninth energy centers, which are located directly above our crown center and relate to our Higher Self. This crown chakra also energizes the pineal gland which governs the brain and nervous system.

The pineal is a sensing gland that transmits messages across the barriers of the body and reaches out to "touch" the senses of other humans without being restricted by distance. For example, if you were in a desolate area by yourself and ten miles away a male human showed up, the pineal gland would sense that there is a person over there. It works much like a magnetic device, to bring people together by sensing where people are, and operating like a periscope, intuiting what is going on around us.

The vital life force cannot permeate throughout the brain and body if there is a serious misalignment or imbalance in the pineal gland. If this occurs, a sense of discomfort in the form of a headache or some other difficulty will manifest. The crown chakra cannot be fully functioning until the root, or first, chakra is balanced. When the energies of the crown chakra and the root chakra are aligned, the Kundalini fire is produced. Once each energy center and gland is awakened and harmonized, the pineal directs the etheric body, including the eighth and ninth centers, to merge with our denser bodies. The crown center is the entrance point into our physical body for our Ascended Self. It is the seat of Divine Will.

RELAXATION & CALMNESS
Removing Resistance

Relaxation & Calmness prepares you to handle both giving and receiving love. Many people feel tense around love, and this can create a built-in resistance to love. For them, love is seen as very conditional and demanding of something in return. True and unconditional love is a very calm and soothing kind of emotion. The frequencies of the Relaxation & Calmness recording can be a good way to become relaxed and receptive to unconditional love.

The Universal Laws of Resistance, as I understand them, include "What we resist, persists... what we resist, we attract... what we resist, we become." Therefore, everything in our lives that is not harmonious is there because we resist it. Whatever the problem—health issues, lack of money, relationship tensions—it usually has resistance behind it. Most often the resistance is not conscious, but it is there. One of the frequencies in the Relaxation & Calmness recording removes resistance and brings peace and letting go. This creates the opportune atmosphere for change.

Change often scares our little self, but our Higher Self embraces it. Fear is what separates the lower self from the Higher Self. Once we remove the fear, we become ONE again and life is filled with JOY. Often we need to destroy the old to be able to create the new. Destruction is only another form of creation. It is merely a matter of viewpoint or perspective.

Many of us are aware that our current DNA structure is not the same as was originally intended or designed. All our DNA strands are still there, but not aligned in a way that can be useful to life. One frequency in the Relaxation & Calmness recording reminds the DNA cells of their original intent by setting up a vibration that acts as a "homing device" which begins to draw the DNA back into alignment. In this way, the DNA will again provide access to all knowledge contained within its structure. This frequency is timely, to say the least.

Also, another frequency in this recording is effective in creating a sense of well-being and helping us reassess and redefine our goals.

LOVE
Abundance of Love

These frequencies are powerful and, as is true for all of the frequencies, it is important that our intentions be clear while listening to **Love**. You may use this recording to feel an abundance of love. It can be an enhancer if you are already feeling loved and loving, but want even more. If you listen with a feeling of anticipation, joy, and wonder and expect to increase your ability to share love, then the vibratory benefits will be great. These frequencies can be felt in all parts of your physical and emotional bodies.

The Love recording is also beneficial for the physical heart muscle itself. The actual physical pumping muscle responds by becoming stronger and more open, just like our feelings. Those parts of our bodies which pulse, such as the heart and arteries, benefit from this recording. The heart and arteries will feel expansive and open and suffer no constraint. This benefit applies to the surrounding lungs as well, and a real feeling of joy spreads throughout the body.

Also, the recording contains a frequency that will give the heart a better balance or equalization with the decision-making brain. Many decisions come solely from the brain, the logical part of ourselves. But, when choosing among options, a collaboration between the brain and the heart is ideal. All things need to work together. A cooperative spirit is a loving spirit, so there is a need for the brain to join in cooperation. That is the intent when the physical body is born. It comes in with a state of balance and a realization that all parts are vital and important. It is only through training that some parts of the body take on more importance. It is we who take the brain out of context and assign it a predominant position.

Unconditional love is the Alpha and Omega, the beginning and the end, the full circle of impersonal love that loves everything, including negativity, with grace and non-judgment. The conditional love we experience here on Earth is the prerequisite training ground, in order that we can clearly see *what we don't want and what doesn't work*. We can then move away from conditional love and practice the unconditional love of the Masters. Then we may see the perfection in everything and know that everything is in Divine Order.

The Love recording can be so moving for some people that they may need to play Relaxation & Calmness first, or at the same time.

HIGHER CONSCIOUSNESS
Raising Consciousness

This recording removes the essence of fear and doubt. It will also "charge up" individuals who are in total denial of their own disharmonies. It will raise vibratory rates into **Higher Consciousness** and enable glimpses of the Oneness of It All.

One of the frequencies acts as a rejuvenator for the soul and brings the soul to its full acceptance of its totality with God and the Universe. These vibrations act as the "glue" for all levels of ourselves to come together as One. The frequencies can help those stuck in third dimensional disharmony to release their habitual emotions.

Two frequencies will raise your awareness to a higher level, while another frequency will bring to the surface the emotional disharmonies that need to be cleared. Because you will then be above the problem, you can see more easily how to handle it and the challenge is not as scary. A good analogy to this would be to imagine you're flying above the heavy traffic of a big city. You see the traffic, but you are not affected by it. You are in a position that allows you to be in power.

Accepting Assistance

Some people, after listening to these frequencies, may require the support of a loved one or friend as they work through these particular "emotional" issues. These frequencies will bring up fears and doubts—often issues that we didn't even know were buried deep within us.

The person who supports us may only need to listen for awhile, as long as needed to be a sounding board. We may find it helpful to discuss the thoughts, feelings and dreams that can arise. If we are able to verbalize what comes up when using the SWE recordings, very often we will hear our own answers. When not allowed to be spoken aloud, feelings can stay stuck and not be transformed as was intended.

This self-disclosure process is the catalyst for a very quick transition. One of the benefits is that it allows us to feel our connection to another person or persons and to know that we are a part of the Universal Self. We then recognize our Oneness and that we were never separate and never could be. There is security in knowing that the idea of separation is a major illusion. Sometimes, to get to that understanding, many old beliefs must come up to be examined and released. The value of a kind and loving ear cannot be exaggerated. However, if we have no one to talk to, we may write down what is coming and look at it from a more detached viewpoint, like an observer.

At the beginning of this chapter, I recommended playing the *Chakra #3* recording with Higher Consciousness. However, the testimonials of Dr. Roy Kupsinel reveal what may happen if we listen to this frequency too much in the beginning. An uncomfortable number of emotions are released simultaneously, creating a great deal of stress if you listen at a rate that is greater than recommended.

At the beginning, it is good to listen to the Higher Consciousness recording each day. Once we can play it for one full hour and no negative emotions come to the surface, we can then increase the listening time to two hours, then three. The idea is to listen enough to bring up issues that need to be dealt with, but not so much as to bring too many things up at one time and create discomfort. If we increase listening time gradually, the progress will be smooth and easy. After awhile, you will be able to listen to it all night and just love it.

BRAIN, COURAGE & PROSPERITY
Accepting Abundance

The **Brain, Courage & Prosperity** recording contains a frequency that will make a minor adjustment to each wave-form in our brain patterns. This will actually place a restraint on poor decision-making and negative thoughts. These frequencies will facilitate the process of changing old beliefs that need to go.

One frequency acts as a rejuvenator of harmony to provide the listener with endurance and the courage to focus on Self. There is a frequency for grounding and another frequency that will help integrate the left and right sides of the brain. One of the frequencies helps us to have better control of our thoughts.

These frequencies encourage the mind to accept the Self in its totality as well as to accept abundance as the natural law of the Universe. By listening to these frequencies, you will have courage to ask for all you need, be grateful for what you have, and recognize your birthright of abundance. The feeling of abundance is enhanced, which is a critical component because none of this can be achieved without having the feeling. As with the manifestation of anything, there must be a connection between thoughts and feelings. No matter how much the brain allows abundance, if the *feeling* of abundance is absent, abundance will not be manifested.

As you will read in the testimonial section of this book, the combination of *Energy Center (Chakra) #6* with the *Mental & Emotional Series* is quite powerful. In addition to greater abundance, I have been so pleased to see how these frequencies have helped those with learning disabilities and attention deficit disorders.

One of the teachings I perceived from my guidance when I produced this series was how important it is to honor the commitment that the soul may have made before incarnating. I believe each of us chose what we need to experience, and if we are not ready to let go of the experience because we have not yet gotten its full benefit, even the SWE frequencies will not eliminate the issue.

I learned this directly when I attempted to help a boy who had dyslexia. I knew from past experience how effective this series can be, and I expected the boy to show dramatic improvement. Instead he showed almost none. When I consulted my guidance, I was told that each soul needs to experience what it agreed to experience and it would have been premature for this boy to eliminate the dyslexia.

This can be applied across the board, regardless of the challenge. If more lessons need to be learned, the soul will not release the issue. This is a personal choice and it is important that parents, teachers and healers recognize and accept it, when the situation arises. We cannot and should not interfere with each individual soul's right to guide itself.

THE SPIRITUAL SERIES

THYMUS
Eternal Youth and Unconditional Love

As you know, the thymus gland is located in the upper chest. In most instances, it shuts down during puberty. Our collective belief in aging and death is really what causes the thymus to atrophy. This gland is thought to be the "Seat of the Soul," and the soul knows that it never dies. It can't.

So why should this gland remain active if we believe that we will die around age seventy-five or so? Because of this kind of thinking, it looks like we simply don't need this gland past puberty and so it just stops working and begins to shrink.

What if we decide that we want to stay young and healthy and just be the joyful being that we were intended to be? If we have already begun the aging process, can we reverse it? The answer is *yes,* if we activate the thymus gland and bring it out of dormancy. By thumping gently in that specific area, using what is affectionately called the Thymus Thump, (Tarzan's favorite!), the vibration will help to activate it.

The same frequency that I used for five minutes in *Energy Center (Chakra) #4,* is combined with the frequency of DHEA, and played for sixty minutes with two pair of elements. I don't like to repeat frequencies, but because of the urgency to develop the thymus, I agreed to do it in this instance. The Thymus recording helps to develop a very powerful immune system. It also opens the way to *unconditional* love for ourselves and others.

About DHEA

Since I began working with the frequencies, every time someone would introduce me to another supplement, I would muscle test myself (kinesiology) to see if I needed it. The answer has been consistently "no," until one day when my friend Dr. Roy Kupsinel introduced me to DHEA. I was surprised to learn that my body said "yes!" After reading all the studies that have been done regarding this wonderful product, I understand why.

Researchers at the University of Wisconsin wrote: "The role DHEA plays in aging and disease is perhaps the greatest discovery of this century." DHEA is the most abundant hormone in the blood stream. Its concentration dramatically increases during puberty and decreases after age twenty-five. It is a precursor of hormones, and to maintain balance, the body will convert DHEA into whatever hormone is needed—like estrogen, progesterone and testosterone.

Even a person in good health needs to develop the thymus. Thus, when I was guided to produce the recording about the thymus, it was natural for me to include the frequency of DHEA. One of my friends who was taking DHEA began to realize that her body couldn't absorb it. However, once she began playing *Thymus*, her body was able to receive the vibration of DHEA, like many other elements. Plus, these frequencies are much more economical, since the DHEA was costing around $30 a month.

TRANSITION
Easing Changes

Transition contains one frequency that deals with acceptance of new things that are on the way during the upcoming time of rapid changes. Another is about honoring of self, using the will to transform energy to a higher frequency moving us courageously into change or transformation. Another frequency is effective in adjusting each person's acceptance of harmony and unification of self with the Almighty.

These frequencies help remove our limitations. Instead of thinking in a little way, we start to think BIG. When we hold beliefs, it's like having ideas in boxes of different sizes. The sides of the boxes are our limitations. The larger boxes relate to the aspects we feel comfortable with, because we trust our capacities to deal with these aspects. The medium sized boxes relate to those areas where we have fear mixed with positive beliefs. But with the aspects where we feel the most fear, the box is so small we can hardly get in—and when we do squeeze in, it is very uncomfortable.

Because we have created all of these boxes, we also have the power to change and eventually dissolve the sides that hem us in. We need a lot of faith to let go of these self-imposed limitations. The unknown, outside the boxes, is scary only when we are trapped inside the box. When we reach out we find Light, Love, Life, Freedom and Joy. It feels so good to remove all these barriers that prevent us from living life fully and freely. To expand, we can gradually enlarge the smaller boxes, and remove one side at a time of the larger ones. My near-death experience helped me to get rid of a lot of boxes. I wish everyone could know what an NDE teaches (without, of course, having to go through the experience).

Belief is really a trap. Often our beliefs come from others and are based on prejudices and preconceptions that we just adopt without questioning. Our beliefs are usually formed at the unconscious or subconscious level. When we are children, in particular, we adopt the beliefs of others like our parents, teachers, relatives and ministers, without conscious decision. If an authority figure says to us, "You are worthless... you are no good," this unsupported "belief" becomes lodged in our subconscious as truth. Once we believe something, it becomes true for us. Often this is the end of looking for a higher or greater truth.

What we believe to be true when we are a child often no longer serves us when we become an adult. For instance, the belief that we should not cross the street certainly needs to be discarded as we mature. Many people believe they have found THE philosophy, THE mate, THE house, THE place, or THE job. It may be true at that time. But times change and so do we, and these changes can signal the appropriateness of letting go of these old beliefs. It would be so great if these beliefs would go the same way they came, with no effort, by just being surrounded with people with other beliefs. Unfortunately, that is not usually the case; the beliefs were created unconsciously, but we need conscious awareness to release them.

Once we remove our limitations, we discover who we really are. What we used to think of as a miracle will become the "normal" way. We'll come to a point where *we won't expect miracles, we'll rely on them!*

The Transition recording will help achieve this. It is very exciting. Are you ready?

MERKABA
Vehicle of Light

A **MerKaBa** is a vehicle of Light around the body. This recording helps create a MerKaBa around ours. MER means Light, KA means Spirit and BA means Body. The MerKaBa is comprised of two counter rotating magnetic fields of Light that contain Spirit and Body. Two interlocked star tetrahedrons that rotate in opposite directions, one with the point up and the other with the point down, make up the MerKaBa. The shape is reminiscent of the Star of David. One star tetrahedron spins clockwise at a certain speed while the other spins counterclockwise at a precise speed. When our body is inside this field, it is surrounded by an impenetrable vortex that extends fifty to sixty feet away from our body. It is literally our Sacred Space where we are totally safe and secure and can readily contact our Higher Self.

One frequency included in the MerKaBa recording creates this spinning effect around the body. The MerKaBa recording is powerful, yet smooth. It can be used while meditating or when we simply want to feel protected against external interference or influence. It is the perfect place to contact our Higher Self. You will experience peace, serenity and meditative states when using it.

ENERGY, DNA & ENZYMES
Harmonious Group Awareness

In this recording, one frequency helps us to know our own Being, know the blood flow process and the process of conscious interaction between the brain and Higher Self. It will help attune us to deeper understandings of these energies. Also it will improve the body's ability to manifest enzymes. As a result, a 30 to 35% compensation can take place in food devoid of enzymes or high in anti-enzymes. Food can be radiated with this frequency and people can be exposed to it as well. It will be especially helpful for those using the macrobiotic dietary principle for healing, in which cooked foods are being utilized extensively. A macrobiotic diet can be a healing regime because of the particular alkalizing effect of the cooked food, but cooking also depletes the enzymes and the deficiency in enzymes stresses the pancreas. Also, through these frequencies, the pancreatic cortex will be strengthened.

On a spiritual level, one will have a deeper understanding of the shift in consciousness that can be manifested internally by brain wave states. As people begin to work more with SWE recordings, a group consciousness will gradually develop. Under such circumstances, a deeper awareness from the group may result. The foremost prerequisite for this consciousness will be the group understanding of the proper utilization of food. The maximizing of enzymes is pivotal to deeper understanding and awareness. Higher levels of interaction with group activities in many different areas will result.

Another frequency appears to bring the energy centers (chakras) two, four, six and eight (the energy center above the head) into focus. This frequency strengthens the digestive system, the connection between the mind and the heart, and the ability to love and to bring love into any situation. The result can be increased and balanced energy in the physical body, particularly the digestive system and the spleen. There may also be improvement in the ability to metabolize various byproducts of the muscular system, like lactic acid (that can then be deposited in the muscles in the lymphatic system). These byproducts can be "reception sites" through which toxic materials, such as bacteria, viruses and others enter the body; this frequency acts to prevent these toxic materials from taking hold. It is also anti-viral, tending to cause the body to release parasites, viro- and micro-organisms. This is an excellent frequency for athletes, because it promotes faster recovery and assists the body in releasing substances that are no longer needed.

Another frequency appears to influence cellular metabolism and the transference of DNA information. This will often involve bringing through an inherited disposition, sometimes attuning us to our grandmother or grandfather or to those of preceding generations. There can be a clearing aspect to this, and a whole new set of energy is released, as if suddenly the ancestors have become our allies. This is an excellent frequency for those who contemplate parenthood, as with this frequency there is an understanding of one's descendants as if seeing the transference in DNA. This can help us select those qualities from both parents we wish to pass on to our child, and attune our DNA to transfer those qualities to the DNA of the future child. When there is deeper understanding of this DNA transformation, we will be able to bring in new information and consciously embody it in the DNA.

On the physical level, this frequency appears to provide some relief for those with auto-immune difficulties such as Lupus. The strengthening of the etheric to the physical connection also appears to take place. It makes sense, when utilizing this frequency, to visualize the changes coming in. In addition to this, there is a sense within the individual of being more than a physical body. The individual may become aware of all the bodies awakening in different areas.

PATIENCE
Connection with Mother Earth and the Plant Kingdom

In the **Patience** recording, one frequency will strengthen those who are taking on new concepts and seeing things in a way they never tried before. This will be an important frequency for travelers because, in their own consciousness, they are making a new connection—a new relationship with Mother Earth herself. Another facet of this frequency concerns the earth's need to establish communication, to blend energy, to bring a sense of empowerment and strength to people, and to release the sadness, the difficulties and the struggles Earth experiences in her relationship with humanity. Understanding the earth as a loving being, therefore as another species, is imperative now. With this frequency, we will gain a deeper understanding of the earth and thus use her energy more appropriately and lovingly.

As we use this frequency, we will feel the power of Earth's energy strengthening all our bodily processes, such as digestion and elimination. This frequency will aid such challenges as chronic fatigue syndrome and fibro-myalgia. Individuals will gain an awareness of the earth's geopathic energy in situations that may be both harmful and beneficial, making this a good frequency for dowsers.

Another frequency relates to people's abilities to receive enlightening consciousness or awareness from the plant kingdom. The Patience recording may allow us a deeper sense of patience, as if we are viewing things through the perspective of the plant. We may come to understand our proper place in various unfolding dramas. Some of us may find ourselves trapped in what appears to be repeating secret dramas. This is caused by our emotional attachment to the attention that other people give us while in the throes of drama. The use of this frequency will be helpful in breaking the attachment to and need for this approval and attention.

There is another direct correlation to the plant kingdom, because without these life giving substances, we could not survive. Yet, the plant kingdom does not seek approval, knowledge or understanding from any of the other beings it benefits or with whom it shares energy.

This frequency assists in the grounding process. There appears to be a physical connection to the minor energy centers, (opening points in the feet), and the ability to be grounded to the earth.

Another frequency appears to stimulate the nervous system so that it functions at a higher vibrational level, enhancing our spiritual capacities and strengthening our ability to process and clear energies such as fear. The lungs also benefit from this frequency.

ONENESS
Communion with All

The **Oneness** recording triggers feelings of infinity and oneness. It contains two frequencies. One frequency enhances the feeling of universal love. The other helps us experience that wonderful feeling of oneness with mankind, the other kingdoms on Earth, and the entire Universe. When these frequencies are used in meditation, people experience communion with all the earth's kingdoms and beyond.

The Oneness frequencies amplify unconditional love.

PURIFICATION
Proper Utilization of Energy

One frequency of the **Purification** recording allows us to take energy in from the environment, not just from food, but especially from air, water and most importantly the helpful energies of the Sun. It also appears to influence the ability to get rid of unconsciously gathered material entities that can interfere with our ability to grow. An example of an entity is someone who has left the earth suddenly and is not aware at the conscious level that he or she is dead. Such an energy, in reaching out to others, can be an unconscious influence. With this frequency, our ability to release such an energy is improved. A higher energetic consciousness will take place and the entity may choose a different way of working.

The natural connection here is through the blood. Many aspects of blood are energized by this frequency. There can be a cleansing and clearing of the blood, plus there can also be an increase of oxygenation by reducing interfering microbes and other harmful bio-organisms in the blood.

This recording is especially helpful to those who have recently been involved in a hospital stay, particularly if they have been unconscious for a period of three minutes or longer. This is the time period in which it is fairly easy for an entity lingering in the hospital to join a body. If one has been involved with surgery, or any hospital stay, this frequency will improve recovery and accelerate healing, as the healing process will not be hampered by an entity attachment.

Another frequency tends to energize master hormones such as DHEA, and seems to improve the brain/body relationship. This is a temporary help for most individuals with Parkinson's. When one utilizes dopamine or other substances, the body continues to compensate. On a temporary basis, this frequency may be useful in improving the physical body process, giving the individual a chance to look at the deeper issues involved.

There appears to be a strong dietary connection between viruses and the overall physical condition of the body. The result of this can be the formation of various molds and fungi in the brain itself, reducing the receptor site attachments and the ability to properly utilize numerous brain chemicals. Viruses have a unique ability to mutate (change form) and hide in places where the blood does not reach. This frequency can help. It will improve the underlying acid/alkaline balance in the body by bringing more healing energy throughout the entire body.

An additional frequency has the ability to pour energy into any given situation. Also, it is helpful for those who wish to increase their focus in a given arena, to increase group activities and to raise levels of understanding in the world. This is made possible by the strengthening of the understanding of proper utilization of energy. This frequency will influence the production and destruction of harmful by-products, such as ATP (*Adenosine Tri-phosphate*), that are utilized in the muscles and then disintegrate. Individuals will have a greater choice as to where their energy is going. This frequency reveals the understanding of the process out of which any sort of doing, work, manifestation etc., occurs. This process moves from the *being* aspect into the *doing* aspect and then into the *having* aspect. There will be a few people who, when receiving this frequency, will experience a sense of lethargy. Should this occur, it would be better to intensify it, relax and do nothing if possible. This will usually pass within half an hour. This is true when working with any of the other SWE recordings as well... instead of resisting, go with it!

GATEWAY
Integrating New Energies

Two frequencies are used in the **Gateway** recording. One frequency allows us to receive the new energies that are coming onto the planet. The other frequency helps us to integrate these new energies. Many people are actually having a difficult time raising their vibration as fast as the planet's. This often manifests as sudden physical disease or dramatic emotional issues coming up.

We need only listen to the news to become bathed in negativity. Reported disasters force us to reconsider our priorities and release attachments. We may resist, but it will become more and more difficult. This recording helps us to revise our values and remember that we have everything we need in order to enjoy our highest good. The secret is not being attached to the material plane. Gateway assists us in releasing attachment and brings an increased sense of inner freedom.

IMMORTALITY
Opening to our Gifts

This recording appears to stimulate a higher endocrine function in the pineal and pituitary. When deficient, these glands will draw on the energy of other glands, particularly the thymus, thyroid and adrenals, creating low energy in the individual. When this frequency is applied, deeper stimulation will occur as if the pineal and pituitary can temporarily come to a state of resonance, finding other ways of drawing in healing.

The primary reason so many individuals have difficulties with the pineal and pituitary resonance is that, in the past, they may have misused mental energy and didn't listen to their heart, resulting in harm and difficulty. So, in this lifetime they have a spiritual filter. They shield themselves from the connection between the pineal and the pituitary, the powerful loving energy, so that this energy won't again become mental in character. This is one reason why individuals on the spiritual path often report fatigue.

It is very important that we become aware of the heart energy.

The highest level of this is an understanding of new patterns, new ways of thinking, and breaking out of old and stuck patterns. Our ability to know the will of the soul and the will of GOD is enhanced.

Another frequency influences cellular regeneration in a youthful manner because it allows greater absorption of Xenon from the earth's atmosphere. Also, new patterns of emotional understanding may be revealed, having a profound effect on the physical healing level in the body. The individual will experience a better understanding of "the many and the One" through inner coordination and inter-dependency, and receive a higher level of loving understanding and connection.

CONNECTIONS
Higher Attunement

This recording contains a powerful frequency for those who wish to understand their own inner soul light and its connection to the Great Central Sun. It is a frequency that proclaims, "I have choices," "I can open up to my own Soul energy," or, "I can use this energy for the gathering of money and power over others." This could be called the clarification of the ego, where the ego takes its proper place. This can be a useful tool for healers as they wish to increase their own healing ability. They will have a sense of their own light moving down through their arms and into their hands. Almost any healing technique will be strengthened by the use of this frequency. The individual, recognizing his or her own inner light, can find new pathways, new ways in which to bring this light outward into the world.

This frequency may draw the attention of the angelic realms, and support a deeper communication with angels. Also, it will help people access a connection to another level—a parallel universe outside the realm of our time and space barriers. An individual may also reach forward or backward in time, once these barriers have been dissolved. It is as if the Being you will be later is loving you now, united with you by the energy of this inner light. Other aspects involve self forgiveness, clarity, awareness, and understanding. There will be a flood of energy for all kinds of healing processes, and the use of energy between a healer and clients, in most circumstances, will be enhanced.

Another frequency enhances the reception of spiritually developmental information and attunes to outside influences from extraterrestrial sources. This also assists individuals who are coming to a more profound spiritual understanding and awareness and yet find themselves blocked by deep fatigue or difficulty attuning to a higher vibration. The inability to receive information will gradually become apparent, and an inner knowing will reveal itself. All one needs to succeed is then present and available.

Another frequency appears to supplement or strengthen our psychic abilities and higher God connection. The connection between the individual and the soul is strengthened, plus a sense of ease and awareness of "the flow" will be experienced.

SHIFTING CONSCIOUSNESS
Love in Action

One of these frequencies has an important connection to all humanity, helping individuals attune to humanity's greater purpose and supporting quantum leaps in consciousness.

Another of these frequencies contains many components that are related to harmonious interaction with other people, as well as to the physical body's immune system. This is a good frequency for shifting consciousness out of humanity's past patterns. However, this can be stressful for those who are unwilling to do this.

This is a good pivotal frequency. People who attune to it will make profound changes, but they may have to stop for awhile, take a break and come back to it again.

These frequencies will allow the larger collective consciousness to become more accessible. This can give the listener deeper insights, connect him with higher levels of energy and a vibration of love with practical earthly applications. Those listening to these frequencies will know how to apply this love in their lives, in their relationships, and in their group interactions in the world. This translates into: LOVE IN ACTION: WILLINGNESS TO BE OF SERVICE. It is an important way in which the larger issue of the development of unconditional love within humanity might move more into conscious awareness. It is the part of unconditional love that is the activation of the will to serve; the choice to make such a step is encouraged by these frequencies. The individual can then see the benefits of such a choice and realize the possibility of higher consciousness, merging of group energy, and the larger picture of humanity. The individual will understand how joyful this can be, and may be more likely to make such a choice because it is *fun*.

This frequency will help open the door to the beauty of love in action.

MANIFESTATION
One Pointed Focus

One frequency in the **Manifestation** recording allows us to connect to our ability to maintain a level of one-point focus. There have been thoughts placed into the DNA intended to reduce our ability to manifest or create a "one-point" focus of concentration (the ability to bring all levels of the various aspects of the brain and mind into harmony). The dissolving of these blocks, through the use of this frequency, will have a variety of effects. The effect will depend on our own willingness to reawaken, to open to our own levels of inner sensitivity and consciousness. Those who are truly working out of joy in their hearts may often find benefits in this materialization process. If this frequency is utilized on a regular basis, we may hold "one-point" concentration as a singular state of consciousness for longer and longer periods of time, whether the frequency is present or not. Viruses held in the spine may also be reduced in their intensity and destroyed by the use of this frequency.

Another frequency is emitted as part of the Solar Spectrum (the energy from the center of the galaxy). This frequency will help to bring in a sense of purpose that will manifest in the physical body. For many, the functioning of the liver will be enhanced. For some individuals who have been on the spiritual path for a long time, and who have a reduced heart energy causing heart murmurs, heart disease or the "backing up" of the arteries, this frequency may be helpful in easing these conditions. What manifests in our world comes from our *being* first, then *doing*, then *having*. This frequency brings a greater understanding of our purpose, an awareness of our "Beingness" and a sense of calm and peace. Also, it has an energy behind it that brings deeper awareness within people and a way in which to manifest it. The results of this shift in consciousness can be an awareness of the role the higher guides play. This can be helpful for group purposes, especially for those individuals who work together for a higher purpose.

THE MENTAL / EMOTIONAL SERIES

This series of recordings is very important, because to be able to achieve our spiritual growth, all the lower emotions need to be dealt with. These recordings will bring our suppressed emotions to the surface.

Once we can listen to the *#21* tape for days without any disturbance, then the *Boomerang* manifestation frequencies can be made available, upon written request (more details follow).

An important thing to remember is that we have arrived at a point where we no longer have to spend years processing. We can ask for GRACE and go on with the discovery of who we really are.

SPACELIGHT
Light Bringer In a Space

SpaceLight contains only one frequency, which brings Light into a space or enclosure. The frequency does not deal with people present in that space, only the space itself. This is similar to the smudging used by Native Americans to clear an area.

The SpaceLight recording is good to use before and after a meeting to clear negative energy. It is excellent to use when you move into a new home or office, or to clear your home or office after a negative person has been there. Negative people may feel uncomfortable in the vibration of SpaceLight and may choose to leave.

I've found that if a person does not have sufficient Light, she or he will be unable to remain in the place that has been "SpaceLighted." Light attracts Light.

The SpaceLight frequency is a good one to play in any place we sense unwelcome discarnate beings. Sometimes when people die under the effects of drugs, while having surgery, or from a sudden trauma such as a car accident, they can become lost and don't know where to go. They may stay in a place where they feel comfortable, like their old house or apartment, which may be where you live now. These confused entities often need help to guide them toward the Light. Loved ones are waiting for them and they simply need to be told to go toward the Light and that they'll be greeted warmly. The SpaceLight frequencies will facilitate their return to the Light. It is best to play the *MerKaBa* recording as well, so that you will be completely protected in your vehicle of Light. This puts you in an ideal place to guide a lost soul back to the Light.

SAFE ENVIRONMENT
Where Heaven and Earth Meet

Safe Environment contains the frequencies of noble gases that were present in the atmosphere a long, long time ago. They help re-establish an environment where all kingdoms on Earth may flourish, and all living things are supported so they can accomplish their Soul Mission.

One of the effects of this frequency is that it will allow the alignment of the third and sixth chakras, or energy centers. When the physical power and the spiritual power are aligned, the person vibrating at a higher level will not be influenced by outside disturbances such as extremely low frequencies coming from electrical transformers or other sources.

CLARITY & FOCUS
Embracing Life Now

The **Clarity & Focus** recording increases inner awareness and inner guidance, because the frequencies are so strongly connected to the opening of the brow center, or sixth chakra. These frequencies help us to receive visual information, to know our purpose here on the planet at this time, and to maintain focus. One of the frequencies has to do with *remaining in the moment*, which is the key to solving many of our issues and being happy, right now. Included is a frequency that helps us see the connection between past lives and our present life, by providing access to information related to a current life situation. The willingness to allow these experiences into our knowingness is a vital link, since the frequencies alone cannot bring these kinds of results. Willingness and intention must be present.

COMMUNICATION
Open to Other Perspectives

In the **Communication** recording, one frequency appears to be associated with the ability to carry burdens and energies of others. This is particularly useful when there is a difficult connection between student and teacher. This frequency may help the teacher to see learning blocks and be able to shift them, to see with the other's perspective, and know the student in a different way. This can result in improved communication and a willingness to let go of the manner in which we see ourselves, which is related to our self esteem. Teachers who are involved in helping students are aware of this, in addition to the learning modality. But, often it is the student who is not willing to change or shift his own vibration in order to receive the new energies.

This frequency helps to establish an inner student/teacher relationship in such a way that the students accept their own messages, understanding, awareness and their own internal experiences more clearly than they had thought possible. This is more difficult than one would think, because in observing our internal experiences, we tend to block, ignore, or deny the experiences that do not always fit with our own larger world view. Both student and teacher will benefit greatly from this connection and attunement.

Another frequency in this recording influences some levels of cellular regeneration, creating a "youthful" effect. This is because of the absorption of greater amounts of Xenon prompted by these frequencies. Also, new patterns of emotional understanding may be revealed so a profound effect at the physical healing level is present. A better understanding of the overall message of cellular regeneration is that, through inner coordination and inter-dependency within our own bodies, we can receive a higher loving understanding and connection with all that exists.

CHOICE
Serving Little Self or Serving Higher Self

The **Choice** recording appears to provide many results, falling in three categories *depending on the orientation of the individual.* Those individuals who are attuned to service to the little self will find that there is a deeper understanding of this relationship and be prompted to a willingness to change. Those who are in service to others will find they have new insights as to how to serve, work with others, and love them. They will instantly attune to a new way of serving, assisting, and loving others that they hadn't known before. In the third group are those in the middle—those unsure of how to proceed. Application of this frequency usually results in some difficulty in attuning to the energies that don't "fit" for them. Thus, it pushes the ones in this "middle ground" to one side or the other.

It is suggested that, during the three-day application of this frequency, there be a reduced exposure to particular frequencies like television. Television seems to push the "middle ground" people in the direction of service to self. For others, it is as if they begin to recognize their inner possibilities and are encouraged to choose their own highest good. This frequency brings in a larger level of the Soul Self, a level of energy that touches the individual so that he may find the path, choose it, and know it.

Another frequency appears to reduce the barriers between many different substances of hormones, neuro-peptides and others of this nature, throughout the body. It will tend to energize the master hormones DHEA and others. It seems to improve the brain / body relationship at the physical level. For those who have Parkinson's, this will be temporary, as when one uses dopamine or other substances, the body continues to compensate. On a temporary basis, this frequency may be useful in improving the physical body process, giving the individual a chance to look at the deeper issues involved. There appears to be an important dietary connection between viruses and the overall physical condition of the body. If there can be an underlying fundamental metabolic change, the body can be shifted into a more alkaline state. As higher levels of oxygenation are produced, the destruction of micro-organisms also results.

MEMORY
Attunement to Many Levels

This frequency brings several levels of attunement. Past life connections, particularly with the animal kingdom, may be increased. In certain individuals, there seems to be some strengthening of their inner animal connection. An individual may listen more easily to the message their animal "totem" may have for them.

There can be an increased awareness of the limbic system, for improved balance across the lateral nodes in the brain, attuning not only right and left, but up and down as well. Connection to the temporal lobes is improved, and awareness of the link between the astral and emotional bodies may be strengthened. Some individuals may experience a heightened dream state as well. This is a helpful frequency to use when we are working with any skin disease, as there can be a karmic connection with a relationship to an animal. Some individuals may see the skin condition worsen if they are unwilling to confront the deeper issues. Some affected by the more widespread skin conditions, particularly those viral conditions in the Herpes or Shingles family, may benefit from this frequency.

In addition, there can be, for some, a deeper connection to their ability to transform energy from one level to another. This will work in the breath, as oxygenation improves. Some individuals, willing to work with such levels of transformation, can switch emotional energy into mental energy for improved memory, recognition, and better learning.

PUBLIC SPEAKING
Getting Out of the Way

Public Speaking helps the individual to receive a new sense of his speaking capacity. An actual shift in the frequency of the voice can take place. The ability to see oneself through the eyes of another is retarded. The ability to transform energy from thought to speech is a process of slowing down the thought rate. In the channeling process, such a frequency is used to synchronize levels of communication. The fear of public speaking is eased, particularly where the individual's connection to those in the audience is part of the problem.

How to apply this recording:

1. Listen to it, bring it into your consciousness, feel it and imagine it.

2. Go on with the speaking.

3. Later, but preferably within twelve hours following your speaking experience, listen to the frequency again, or imagine it, once again in a meditative state. You may at this time receive additional messages, or you may feel energy from other occasions when your guides or your Higher Self have helped you. To complete the cycle, you may experience energies from everyone in the audience as this process comes full circle.

This frequency may influence your ability to hear your guides and to recognize your influence in the Universe. Those who meditate truly benefit from this frequency.

BODY WISDOM
Body Total Self-Acceptance

Body Wisdom helps us realize that we have based our lives on certain beliefs and patterns that no longer serve us, and that we must change one way or another. The belief that women must adhere to a certain body size standard in order to be loved and accepted, is a pattern that needs to change. Women in general suffer from self-hatred and self-rejection of their own bodies, because they may not comply with the unrealistic standards a male-dominated society has placed upon them. Men, too, have body image concerns, such as balding, height (too short or too tall), and others. Both men and women have allowed this to happen, so this has been a situation that both have created. Love is always at the center of any life enriching focus, so love for ourselves and our bodies—just the way they are—is where we begin.

One of the frequencies works with self-acceptance and changing our beliefs regarding what is expected of us by society. Most of the discomfort and problems we face come from faulty belief systems. Whether for a woman it is to look a certain way, or for a man it is to be tough, this frequency addresses these old belief systems. It is extremely helpful to shift our focus away from what is bad and wrong about our body, to what is good and right about our body, today, right now! We tend to focus on the imperfections and forget all the benefits our body gives us each day. This approach will help us see ourselves as 90% perfect, rather than focusing on the 10% that is not quite perfect. These frequencies help us by supporting positive changes and enabling us to make the changes more easily, because now we can notice all the good things about our body. One of the ways we can accomplish this is to do a daily appreciation and acknowledgment meditation which will shift our focus and create a genuine love for our body, just the way it is.

Daily Appreciation & Acknowledgment Meditation

We call this process "Body Love." Spend five to ten minutes each day doing this meditation. Begin with your toes and work your way up your body, part by part. Speaking aloud (it is very important to speak aloud in order to change the pattern in your ethereal body, and to put love vibrations in your aura), give love and appreciation to that particular body part, and then consciously focus on what that body part has done for you today. Here are some examples:

"I love my toes. I appreciate and acknowledge my toes for the wonderful service they provided for me by helping me

*stand and remain balanced while I was at the bank today.
I love my ankles. I appreciate and acknowledge my ankles
for providing such needed support for my legs as I went dancing tonight. I love my calves. I appreciate and acknowledge
my calves for giving me such strong support today and allowing me the privilege of walking to my daughter's grade
school and bringing her home safely. I love my knees. I appreciate and acknowledge my knees for giving me the flexibility to play tennis today with my friends."*

You see what I mean. When you become specific about thanking
your arms for enabling you to hug and embrace your mother today,
or hold a favorite book up for you to read today, or whatever, it
brings the love, appreciation and acknowledgment into concrete
reality and enables you to focus on real life. These are meaningful
events for which you can genuinely thank your body, regardless of
its shape or size. When we really begin loving ourselves, we don't
worry so much about others loving us. It is no longer necessary that
everyone loves us. This allows us to be more open, vulnerable and
much less in need of the protection that excess weight provides.

In order to transform our body, we must change our belief
system. This frequency assists in that meaningful and supportive
change. Play Body Wisdom in conjunction with *Addiction, Relaxation & Calmness, Energy Center (Chakra) #4,* and *Love.* You may
play three CDs or tapes simultaneously or merely rotate them.

ADDICTION
Be Really Happy Again

All of us have addictions. Some are addicted to substances such as
drugs, alcohol, cigarettes, sugar or food. However, this type of addiction accounts for only 30% of all addictions. We can be addicted
to power, control, drama, abuse, being right, being on time, and
countless other ideas, people, behaviors and patterns.

Addictions, or emotionally charged desires, not only diminish
our daily happiness, but ruin our life with negative emotions. Feeling upset because someone is ten minutes late, wanting to control

everyone and everything around us, worrying over what is to come, piling up money for a future emergency, being horrified every time an ounce of uninvited fat appears, these are only a few symptoms of addictive behavior. Basically, anything that is out of balance and makes us less than happy is an addiction. But how do we get rid of these addictions? With these frequencies, we have to do some work. When we begin to work on our addictions, it is important to remove all judgment of ourselves and to work to achieve unconditional self love.

First we must acknowledge the existence of an addiction. Second, we must have a strong and sincere desire to release addictions that have never brought any positive results in our lives, and third, we must transform our other addictions to preferences.

To begin, make a list of everything you are addicted to. Use that list to fill in the blanks in the intention statement that follows. Write out your intention and speak it aloud before playing the Addictions frequencies:

> *"As I absorb the frequencies of the Addiction recording, I easily release the following addictions:*
> _____*and* _____*."*
> *"In addition, I transform the following addictions to preferences:* _____*and* _____*."*
>
> *"I am a powerful Spiritual Being who easily releases addictions. I am free of addictions because I choose to be."*

It helps also to play the following recordings: *Relaxation & Calmness, Energy Center (Chakra) #4, Love,* and *Connection.* This combination helps us forgive and understand ourselves while seeing ourselves through God's eyes—as lovable, precious beings just the way we are, addictions and all. Once we accept and love ourselves, change is easily made.

The Addictions frequencies will:

- Help release recognized addictions as well as others we may not be aware of.
- Help remove cravings for substances at the cellular level.
- Fill in the space with a frequency that sustains life and harmony in the cells.
- Release and transform addictive patterns.

When we are free of addictions, our actions will be characterized by wisdom and oneness. Our old unconscious reactions to life become a conscious response based on love. In the end we will be living a happier life!

EMOTIONAL RELEASE
It is OK to Cry

One of the **Emotional Release** frequencies appears to modulate pain reaction, and may, under certain circumstances, alleviate it by shifting the burden and load of excessive nervous impulses from one set of nerves to another. But, more importantly, the underlying cause becomes apparent. Stimulation of the higher functions occurs, so that the individual can alter the circumstances that produce the difficulty or pain. Also, this frequency appears to stimulate certain homeopathic principles that are associated with the release of pain. It will accelerate the process by which healing energy, at a higher level, makes its way deeper into the individual. There may also be a strong emotional component associated with this. The person may experience outbursts of tears, or a spontaneous release of emotional pain. This can be beneficial for people who don't cry. With this frequency, we can gradually open up and let the natural process of crying release what needs to come out.

Another frequency appears to move energy vertically, contributing to the overall balance of the *entire* energy center (or chakra) system. When balance in this system is achieved, the individual experiences internal harmony and tends to find balance and harmony in groups. When in pain, attention is led to refocus, moderating the pain.

Another frequency appears to assist in the production of endorphins in the brain, thereby helping those seeking an inner state of peace.

BRAIN / BODY
Walking our Talk

This frequency brings a shift in brain chemistry that could be helpful for some symptoms of Parkinson's. The etheric body is assisted so that it may more easily work with the physical body, resulting in the ability of dopamine receptors in the brain itself to metabolize and work with various substances that are eventually necessary throughout the body. These same energies usually tend to increase the adrenal cells' ability to work, and thus unconsciously provide the action of physical movement, even without direction from the brain.

This deeper attunement between the physical and mental is assisted by this frequency, and thus, for many individuals with Parkinson's, symptoms may be reduced. We must also understand that there are important underlying capacities associated with this disease that must also be looked at and treated. The most important of these is the proper utilization of food in the body, because poor dietary patterns will certainly increase the risk of Parkinson's. The more we deny our own blocked patterns, judgments, inability to emote or allow powerful energies through, the more we may be associated with this disease. Indeed, this frequency can be helpful in shifting some of these energies, giving us different points of view about our blocks and sometimes enabling a permanent shift. However, there will be some individuals for which this frequency will have no effect at all. These individuals may have a deep inner sense that they need to withdraw from the world.

The *Safe Environment* recording can be accompanied by the Brain/Body for better results.

Another frequency in this recording works with the underlying condition, which is based on the body being way out of balance in regard to acid/alkaline levels. This is directly due to the ingestion of food and the utilization of drugs, antibiotics in particular. Using this frequency, those with fibro-myalgia will be benefited by a shift in their awareness and consciousness, so that it is easier for them to be open to new dietary principles—new ways in which they can eat more appropriately. Also, some of the more difficult yeast based organisms will break up.

TAPE # 21
Light Bearer

The **#21** recording contains a frequency that holds "new" Light and well-being and positively influences each cell. It is a beneficial frequency to use as a way of focusing on the *understanding of truth*. Darkness resists this frequency.

I was guided to create #21 to replace a three-hour custom session I previously facilitated for clients. Many people who participated in the custom sessions told me that each session helped them to become aware of issues that prevented them from raising their consciousness. These sessions had to be repeated at various intervals in order to keep the progress going while each client focused on his or her inner work. It was an effective process, but I searched for a frequency that could replace the long, repetitive sessions. I wanted to broaden the range of people who benefit from this process without their having to go through the time and expense to be here in person. This tape was the answer.

The #21 recording is very powerful, so it is important that it be used properly. I urge you to *begin cautiously.* Start with a clear intention and then listen for no more than two to three minutes at a time the first day. Some people can listen to the entire sixty minutes after only two weeks. Others must remain at a maximum of twenty minutes a day, even after listening for a year. The key is to tune in, be aware of your own response, and honor it.

It is important to allow processing time to work through, understand, and release each issue as it comes up. By doing it in a gentle way, the process can be comfortable and will not disturb you. If you ignore this warning and play the tape too long too fast, it can trigger a healing crisis—which isn't bad, but can be pretty uncomfortable (see Chapter 10).

Remember, this is a very powerful frequency! Eventually, you will achieve the objective of being able to listen to this tape all night long and for several days. The #21 recording brings much Light and great Joy.

THE PHYSICAL SERIES

PRANA
Life Force

Most people benefit from the oxygen frequency in the **Prana** recording. Oxygen is the element that influences all life forms and substances. We cannot survive long without it. Oxygen is so basic to physical existence that, to explain the working of the body, we must start with the cell. Oxygen is an essential element of the primary double helix structure of DNA and RNA molecules. Breathing with continual inflow and outflow of oxygen molecules, through the lungs, is essential in order to provide oxygen to the entire body.

Oxygen is one of the most abundant elements on earth, and hydrogen is the principal element that balances oxygen. When there is not enough oxygen in one area of the body, it is generally due to too much hydrogen. This situation creates pressure somewhere in the body. Perhaps the pressure is within blood vessels or on the skin or even in organs. If they swell, discomfort results. For example, pressure can manifest as a headache. Breathing deeply will help relieve such pain and pressure.

The opposite of too much oxygen and not enough hydrogen is experienced as a spacy or dizzy feeling, which contrasts to the feeling of pressure.

When we express emotion, we automatically breathe faster, which in turn brings more oxygen into the body. The prana recording assists in emotional clearing by providing more oxygen to the body.

People with respiratory difficulties caused by body imbalances or high altitude experience more energy and balance when using the Prana recording. Singers are able to perform better, even in a high altitude. Others use it while driving in their car, if they feel tired. The Prana recording can be played during seminars or afternoon work sessions, to stave off the sleepy feeling we so often get in the mid to late afternoon.

Many people have difficulty concentrating after a meal because digestion takes a lot of energy, and leaves them feeling drained and sleepy. Little energy is left for thinking. Prana is a good remedy.

Playing the Prana recording before or during exercise will enhance stamina and speed. You may run faster than usual and you will also be able to go longer.

HAIR
(On the Head)

One frequency of the **Hair** recording appears to stimulate the body's ability to transfer energy through the spine, and therefore through the energy centers, in a vertical direction. An example of this might be a male that is experiencing baldness due to a high level of testosterone. There is a need to re-circulate the sexual energy, move it where it is appropriately felt, and balance it. It also affects the individual's energy circulation while in groups. When a person feels balanced within, getting along, understanding others and working together in groups becomes appropriate and harmonious.

Another frequency appears to stimulate circulation in the skin, particularly in the area of the scalp. The hair that was lost the last will re-grow first. Also, many people notice that their hair grows faster. Often, there is an imbalance in the hormonal chain, so we added the frequency of DHEA. The DHEA is a precursor of hormones and supports the glands by producing the hormones in a balanced way. This frequency can increase the body's ability to recognize the need for proper balance throughout the entire hormonal system.

CIRCULATION
Cleansing and Energizing

Circulation assists with improved circulation of both blood and lymph. It works to cleanse and energize the arteries, veins and lymphatic system. Essentially there are two primary benefits of these frequencies: 1) they open up the arteries and veins which carry nutrients into the body and toxins out of the body, and 2) they stimulate the lymphatic system and help it fulfil its important function of keeping the body well and guarding against invasions from foreign bodies.

These frequencies provide rejuvenation of the circulatory system, and in turn, benefit the skin and hair. Increased energy is a by-product, because better circulation creates a desire for more physical activity. When physical activity increases, cellulite reduction is a likely result.

A number of spiritual counselors and healers, who use the SWE frequencies with their clients, have shared the value of these frequencies when dealing with unwanted entities and implants. The Circulation recording will not remove a physical implant in the body, but it will help neutralize the influence of the implant, if that is the client's intention. This frequency assists an individual's ability to resonate so strongly within themselves that intrusions from foreign bodies will not be allowed.

CELL REJUVENATION
Reawakening the Blueprint

In order to rejuvenate the cells, a prior knowledge of the "ideal" is implied. Rejuvenation means restoring the natural, normal, youthful and healthy state that each cell originally enjoyed. Each rejuvenated cell remembers its original blueprint that directs each subatomic part of the cell to do what it was intended to do.

Cell Rejuvenation includes a frequency designed to unite each cell in harmony with all the other cells and to support a process of harmonious relationship within itself. From its inception, each cell understood what its function was to be and how it was to gain its nourishment and growth. Somewhere along the line, however, some cells lose their memory. One frequency in this recording reminds the cells to awaken and reminds them of what it was that they were intended to do. It's as if the cells have fallen asleep and have not been creating and maintaining a climate within which they could prosper. By neglecting their duties, they create a climate in which foreign bodies and parasites can enter and "settle in." By rejuvenating the cells, an environment is created that fosters growth and prosperity for the host, instead of growth for foreign bodies and parasites.

Another frequency helps to create an environment where unhealthy cells will be unable to grow or reproduce. As a result, these unhealthy cells vanish from lack of food. Cells that do not resonate with the frequency of well-being will remove themselves, because they are unable to reside in a climate that does not resonate with their frequencies.

Along with Cell Rejuvenation, I recommend playing the energy center (chakra) frequencies appropriate to the area, plus *Relaxation & Calmness*, for maximum benefit.

BONES
The Joyful Now

The **Bones** recording contains two potent frequencies. One has to do with remaining in the moment, the Now. The other addresses the bones themselves. The frequency will entrain the bones to vibrate at their ideal rate. The bones are like a factory where blood cells are produced in the bone marrow. If a person remains in the moment, then blood production within the bone marrow will be of better quality.

Many people experience aching joints, or their bones becoming brittle. These situations relate strongly to emotions that are usually connected to the past. Sometimes anxiety about the future may be involved, but almost always feelings of guilt, fear, inadequacy or insecurity play a key role. All of these negative emotions can drag negativity from the past into the present. Naturally this negativity has an effect on the physical body, creating an imbalance within the bones which prevents them from doing their job.

When one is focused on the *now* and not the past or the future, it is possible to express happiness, joy and abundance because the entire feeling and climate of the physical body is in sync. Each cell, membrane and part goes about its task, basking in the feeling of joy and well-being and knowing that all is well. As a result, everything works together smoothly and in harmony.

It is a very positive thing to function from the now. The innate feelings of joy and abundance are projected into your future *and* your past, thereby pulling into your present all experiences which resonate with the feelings of abundance and well-being.

HEARING
Inner and Outer Clarity

Hearing contains several different frequencies which relate to improved hearing. One frequency normalizes the inner ear with balance and joy, thus affecting whatever might be causing hearing disturbance. Another frequency helps with problems related to infection, and potentially, the correction of hearing loss. You'll notice when your outer and inner hearing apparatus are balanced, you'll experience a feeling of well-being.

Additionally, one frequency has the ability to clarify inner and outer hearing if this is the intention of the listener. If played with *Energy Center (Chakra) #6*, the hearing of inner guidance is enhanced.

HYDRATION
Cellular Connection

One frequency on the **Hydration** recording appears to allow the body to balance the energy moving up and down, allowing the Kundalini energy arising in the spine to move properly. It seems to have a connection to the relationship between oxygen and hydrogen by increasing the absorption of water in the body, as well as enhancing the absorption of nutrients from other liquid substances on the cellular level. The cells will become more discriminating, taking only what is nutritious and rejecting that which is not.

Another frequency will deepen the sense of internal revelation. The individual will understand concepts that were previously unclear. An increase in connection among the cellular structure, the soul and the brain will occur. It appears that the conduction of water through the connective tissue will increase. The kidneys will benefit by this stimulation. The condition of the skin is also greatly improved.

COMPASSIONATE HEART
Happy Heart

One frequency in this recording assists the heart and what it stands for. One of the benefits is the ability to know and accept love, and to give it freely without expecting or needing anything in return. On the physical level, this frequency appears to influence absorption of Vitamin C and, to a smaller extent, Vitamin C production. The human body has the capacity to produce Vitamin C; however, this has been lost through genetic changes, many related to humanity's survival. Women generally produce much more Vitamin C than men. This is one reason why men have more heart problems than women, particularly in the arteries around the heart. Vitamin C is essential for flexibility of the arteries, and when we have a sufficient quantity, we will usually avoid most heart difficulties.

This frequency also appears to allow the pericardium to shift energy in order to make transfers from one level to another. There are important nerves all around the heart, and they may improve in their ability to transfer many kinds of energy in and out. This frequency appears to provide a balancing of the magnetic and electrical nervous impulses.

Another frequency relates on a higher level to our ability to constantly take in and release all ingredients necessary for life. It strengthen the entire body by increasing it's ability to move oxygen in and out of the cells.

Another frequency appears to stimulate the body's ability to transfer energy through the spine and into the energy centers. Each gland produces its own hormones, and each gland needs to be balanced to allow the body to function optimally. At the higher level, this frequency affects the individual's energy in group situations. When a person feels balanced within, mutual understanding, companionship, and cooperation become easy and pleasant.

There may also be an inner revelation regarding the choices an individual makes. The individual may come to the realization that his earlier actions have not been beneficial to him.

PANCREAS
Reprogram

The **Pancreas** frequency works from the outside and from the inside. First, we want to listen to it and receive the frequencies for our body. Second, we need to play it in front of a container of water. Spring water or filtered water will do. In effect, we will be charging the water with the Pancreas frequency. Place the player device within ten feet of your drinking water. Play the Pancreas frequency for one hour when charging the water. Drink 40 ounces of charged water daily.

Most of us had an excessive amount of sugar intake when we were young, and perhaps even up until today. This frequency enables the pancreatic function to return to normal and do its job.

You will enter into a period of much greater self-acceptance and self-love. Based on that love, your body will start to reprogram itself. Your pancreas will function ideally. This will enable you to make the changes in your life that you would like to, including becoming more healthy and physically fit. Remember, what you focus on expands. Put your focus where it can be the most helpful to you. Focus on your body and how wonderfully it is serving you, right now.

PERFECT VEINS
Receiving and Giving

One of the frequencies on the **Perfect Veins** recording appears to give the nervous system the ability to influence the blood flow through the veins and the body, not merely around the heart. It seems to aid in the absorption of Taurine and Lysine, two amino acids included in the recording. These frequencies work to repair the veins when under stress. At the symbolic level, these frequencies rekindle our awareness of those aspects in our world that have been valuable to us, which need to be re-circulated. This recreates and enables us to project these precious gifts into the world.

Another frequency relates to the individual's ability to constantly take in and release all that is necessary for life and strengthen the body's ability to move oxygen in and out of the cells. This frequency also relates to love components, through which the individual is able to love appropriately by giving and receiving love and letting go.

STABILITY & STILLNESS
Perfect Coherence

One frequency in the **Stability & Stillness** recording appears to work with the temporal lobe and its connection throughout the entire body and heart. It lines up the otherwise incoherent frequency associated with the heart. The coherence effect of the heart can also be enhanced. However, an individual who has a heart problem may not be willing to accept or to reach for that coherence. This is why it is important for the individual to be familiar with the effects this frequency appears to have.

Another frequency appears to be helpful for the individual who is seeking an inner state of peace by increasing the production of endorphins in the brain.

VITALITY, NEW HEALTH
Let Go the Old, Welcome the New

One frequency in this recording will enhance the manner in which people absorb new information, visualize things in a new way, and hold this new vision in their consciousness as something useful to them. It will also allow the release of several levels of emotions, especially regarding toxic material found in food. Whether improperly stored food is processed or uncooked, natural or artificial, it moves through the body causing stress, particularly in the small intestine. This frequency appears to allow the release of stored energy associated with the entire process of absorption. Individuals involved in dietary change, when seeking a pure diet, will also benefit from this frequency. It appears to have the ability to enhance and strengthen the small intestine.

Another frequency appears to have a component that has to do with the matter of selective elimination of any small components in our lives that no longer serve us. These can be in our relationships, the way we organize our lives, or little habits that annoy others. Our ability to perceive and make the shift regarding these will improve. This frequency also acts as a correcting measure for the immune system, as it works on many different levels. One of the first lines of defense is the flora of the small intestine. For many individuals, it has been significantly compromised. The T cells are felt to be the body's last line of defense. This frequency will have a tendency to address whatever areas are necessary to be strengthened, and for many, the only area that can be strengthened is the T cells.

Another frequency occurs at a higher level and relates to the individual's ability to constantly take in and release all components necessary to sustain life. It will also relate to love components which will enable the individual to love appropriately, *giving and receiving love and letting go.* Another aspect of this frequency appears to strengthen the entire body by improving its ability to move oxygen in and out of the cells. It also appears to create, from raw materials, some of the necessary components, particularly hydrochloric acid, which must be found in the stomach for good digestion to occur.

VERY SPECIAL FREQUENCIES

BOOMERANG
Manifestation

When I began this work, I was given a frequency that accelerates the manifestation of our thoughts. I was told to use it very carefully indeed. I named these frequencies **Boomerang** because what we think will manifest faster than normal, and so our thoughts will boomerang back to us as manifestation.

I am not allowed to sell these frequencies and I'm certain you'll understand why. If these frequencies fall into the hands of people full of fear and negativity, the frequencies will accelerate those thoughts and feelings. This, needless to say, would not be the best thing for the world or for the person! I am only permitted to loan it for a short period of time and then the person must return the Boomerang recording to me. This, of course, is another reason for the title.

Everything is happening much faster than ever before, because of the increased speed of all the energies coming to Earth. In fact, we can more easily observe our own thoughts because we manifest what we think more quickly. It is very important that the primary purpose for wanting to use these frequencies is **spiritual in nature**, although the actual manifestation may result in any number of ways, according to what we need for our progress.

At present, for instance, one might think, "I want to connect with my higher guidance." However, it may take awhile before this connection occurs. After using the Boomerang recording, the connection can be expected, possibly within the same week. It will be smooth, easy and natural.

An example of results received by one of my clients is the materialization of a personal relationship that had been desired for a long time, but had not happened. Until the Boomerang was played, that is.

I personally noticed that even after idle thoughts of wanting to talk to someone or needing to get or do something, the telephone rings with the answer! This has effortlessly helped to smooth my life. There is more magic everywhere.

Once a person has worked with the Foundation Series and has been able to listen to *#21* (Light Bearer) and *Higher Consciousness* for several days without a flare up of uncomfortable physical or emotional symptoms, they then can arrange to use the Boomerang recording. Because I am not permitted to sell this tape, a donation for the Center would be very welcome!

How do you arrange to listen to the Boomerang recording? First, you should write to me describing what you want to manifest and why. You keep a copy of your letter. Second, after receiving the recording, you will need to arrange an hour and a half of uninterrupted privacy to meditate, reflect on your intentions and review your letter while you listen to the tape.

It is critically important to think only positive thoughts while listening to this recording! When you accelerate manifestation, you certainly do not want to accelerate manifesting the negative. We must be very careful with this recording.

It is usually enough to play the recording once all the way through, because *the increased ability to manifest thought will continue.* Some people feel they want to listen to it a second time. This is fine, but you will not get additional benefits if you listen more than twice. After this, Boomerang must be returned to me immediately.

LEAN BODY TAPE

Because of the situation I had with osteoporosis when I was in my twenties, I have been careful with my weight, as three pounds more than my ideal weight created pain in my back. I also had a high level of cholesterol. I found that the Atkins diet not only reduced the cholesterol, but also kept my weight stabilized. When I moved to the United States in 1984, I used a magnetic pad for five years to alleviate the pain in my back, and during that five years I gained around ten pounds! The arthritis and osteoporosis were still present but at least, due to the magnets, my back was pain free when I got up in the morning. I still had to be careful with my body; I couldn't sneeze, dance or move rapidly.

I tried to go back to the Atkins diet to lose the extra pounds, but it did no good. I wondered why the Atkins diet worked for me in Canada and didn't work in the States.

Our parents ate fat; they did not have any of the diet food we feed ourselves, and the overweight condition so common now, was different then. When cows were fed natural grass, they produced a lot of conjugated linoleic acid (CLA). Some studies have shown that

CLA reduces body fat. The main source of CLA is beef and milk. The percentage of CLA in whole milk in 1963 was 2.81%. By 1992, the percentage of CLA rarely exceeded 1%. Is it possible that our milk and beef does not give us what we need? Maybe the recommendation to not eat too much beef and milk products because of saturated fat is appropriate because they no longer provide the CLA so necessary for the transformation of fat into muscle and energy!

After I began using the SWE frequencies, the osteoporosis reversed and the arthritis disappeared! For the last six years, I felt so great, not taking any supplements and not having any thoughts about my bones hurting, that I did not pay *too much* attention to my body weight. When I tried other diets, the results were contrary to the expectations. It was difficult to lose a few pounds and I gained more when I stopped dieting, even if I ate less than before. Because I had so little success keeping my weight down, I learned to love and accept my body just the way it was—fat, bumps, and all. I could at last sneeze, dance, carry heavy things and act as if I was 30 years old! It was not perfect, but was much better than the pain and lack of freedom.

Since the pain that I endured for many years was gone, I felt I was in great shape. Now that I have gained over twenty-five pounds without pain in my back, it is time for me to complete the test!

I came to a realization about my thyroid. I have been so concerned with the problem my son Robert had, when his glandular system shut down at age five, that I didn't realize I had a problem too. Only recently I made the connection between my body temperature (around 97.3 instead of 98.6), with the possibility that my thyroid gland was not performing ideally.

When I used kinesiology to ask my body the percentage of efficiency of my thyroid, I was very surprised and amazed to receive only a 60% efficiency, because I felt on top of the world! What will it be when I achieve 100%? Also, if I look at my productivity and my energy level, I always perceive myself as having a lot of energy, so I never question it. This attitude of not asking about my own state of wellness was a great lesson for me. We have a tendency to see what another person may need, but too often we don't realize what *we* need. I guess it is why we have family and friends; they help us to see our blind spot.

The thyroid connects the glands located in the lower part of the body to the glands located in the upper part. The thyroid signals the pituitary and the hypothalamus of the situation in the body. The thyroid gland has a very important role, which is to command how much energy will be used for the internal functions of the body, and how much energy will be used for the external activities like work and play.

The thyroid is the controller of metabolism, which determines at what speed and in what quantity the body will utilize the food taken in, to provide energy or create an energy reserve. This gland has been programmed to analyze situations regarding the quantity of food we take in and how it will be spent. For instance, in the past, there may have been times when food was not accessible and the body had to adjust pretty quickly to deprivation. It was and still is the role of the thyroid to give the body the signals that will prevent the spending of too much energy in order for the body to make some reserves in case the lack of food lasts a long time. So the thyroid will simply provide less energy for external work and play, and transform the food into fat that can be used later if the lack of food persists. This mechanism was very important for survival.

The conversion of carbohydrates, proteins and fat into glucose is one of the numerous roles of the thyroid. As we age, our ability to metabolize glucose is reduced, which causes aging, often with degenerative diseases and weight gain.

These realizations pushed me to go a little bit farther in my research and create a recording that will help activate and assimilate CLA, the thyroid hormone, chromium with zinc, and the magnesium with sulfur.

The **Lean Body** recording appears to normalize the thyroid gland. If our thyroid is producing too much or not enough of the hormone, the recording will bring it into the ideal vibration.

The first day I listened to the Lean Body frequencies, I couldn't believe how much energy I felt. I realized that the energy I *thought*

I had was coming only partially from my body. The rest was coming from my will. I guess I have a strong determination to accomplish my mission and my commitment drives my life. Within a week of listening to the Lean Body recording, my thyroid was at 100% efficiency. I started eating twice as much and making the best choices in my food. Because I have more energy, I can accomplish more in less time, so I have time to prepare good meals for myself. My thyroid is providing all the energy I need because of all the food I give her; no scarcity anymore. At the same time I am transforming my reserves (fat cells) into energy, so I can wear clothes that would not fit six months ago! What I realized is that usually, at 10 P.M, I would stop everything and go to bed. Now, I can choose to do another project without effort. I understand why I had procrastinated on certain things. Not anymore. I have plenty of energy and also I have my will. It is abundance, and I love it!

Can you imagine? No diet, no pills, just playing tapes that will transform the fat into lean tissue, while giving us a lot of energy. When we are fully energized, the sky is the limit, effortlessly.

One of the side effects of this tape will be rejuvenation. I am committed to bring my body into a state of perfection.

When I was 51 years old, the overall condition of wellness in my body was around 61%. For the last six years, I created SWE tapes and built the company to its current size of over seventy people. Besides the work and travel that took all my time, seven days a week, I had been listening to SWE tapes whenever I was home. Today, at 57 years old, the condition of my body, mind, and spirit has reached a level of wellness of 79%. I continue to improve as I become more aware. My desire to live for hundreds of years is becoming more real.

Before, I just talked about it, and used the excuse of lack of time to skip exercise and prepare good food for myself. No more. I have all the time I need to do everything I plan while I am using SWE tapes. I make better choices because I have more energy. This extra energy gives me more time to do research, to exercise, to listen to my own body wisdom in order to eat the perfect food.

The Lean Body recording is available now. You can order it with the Foundation Series.

WHAT'S NEXT??

I have more ideas about other recordings that will make the process of rejuvenation faster.

One idea concerns my own condition—the numerous bumps (calcium deposits in the soft tissue) that I acquired on my body starting in 1983. Since I began to listen to the tapes in 1992, no more bumps have appeared, but the existing ones stayed there. I was so happy that the arthritis and osteoporosis were cleared up that I didn't pay much attention to my bumps, since they were not painful and not life threatening. However, when I seriously started thinking of the implications of living a long time, I decided that I had no choice but to resolve all the little things that were not functioning perfectly.

Joseph Scogna, in his body of research, attributed to the parathyroid gland the role of controlling the use of calcium in our body. I discovered that *my* parathyroid gland was the main culprit regarding the way I assimilated calcium. Because of the radiation I received earlier in my life, my parathyroid was severely damaged, and was unable to control the hardness of my bones and teeth and the softness of the soft tissue. My body was "misdirecting" calcium from my bones to my soft tissue! I have created a tape that will work on this and am experiencing it now. It will probably be available within a few months.

By the way, I didn't take a bone density test, and don't plan to have one, I feel I have received enough radiation in my life.

Testimonials

The following testimonials are from users of the **SWE** tapes. They are not presented as scientific data, but as real witness by real people. These true stories are not prioritized in any particular way.

General

Janet Keeney, Deacon
First Congregational Church, Wilton, CT
What I noticed in the last six weeks, since using the SWE tapes, is that my body and hands used to shake uncontrollably. I could hardly hold anything. Now I hardly shake at all! I can hold a glass of water or a book without spilling or dropping it.

My husband has noticed an over all sense of well being in me. I am calmer, happier and am able to handle experiences in my life with a new maturity. I am making better choices for myself and receiving with joy. He said to me, "I want what you have – how did you get it?" I can hardly wait to purchase more tapes.

Carol Miscio & Darshan Singh Khalsa, Reston, VA.

We first experienced Sound Wave Energy in February, 1998; it was Carol's birthday and Darshan had just given her a good quality sound system. We heard our first Sound Wave Energy tape later that day and immediately knew that this was a tool for transformation (and why she had received the sound system). We took the STAR Training in March, 1998. The immediate noticeable effects of regular listening to the SWE tapes were the following: For Carol the immediate effects were of emotional release, crying without knowing the reason why. Later effects included a feeling of being much more grounded and centered. Many noticed that Carol's intense energy had become much softer and gentler. For Darshan, an outburst of creative energy. He went through a period of needing only two or three hours of sleep a day and using the rest of the time to write, draw, sing, dance and meditate. Darshan noticed that his vocal range had increased by a few notes each way and a general feeling of being more "solid" or in the body, as if each cell somehow had more depth and mass. Now, a few months later, sleep patterns have returned more to normal, but the other effects are continuing, giving Darshan the feeling of a deep reworking of his entire body.

We love listening to the tapes and sharing them with others, and have set up auto-reverse systems with sub-woofers both at home and at work. The esoteric traditions have long predicted that the healing modalities of the future will primarily use sound and light vibrations. We believe that Sound Wave Energy brings this vibrational healing technology to the planet in a way that is easily and reliably replicated so that it can have effects on masses of people. Carol's birthday gift turned out to be a gift for all we meet, a true tool for transformation. We feel that sharing these frequencies is a way to be of service to the entire planet.

Dr. Katharine More, Chiropractor, Dana Point, CA.

The SWE tapes are powerful tools for change. Shortly after I first started listening to the tapes, I experienced an explosion of creativity. Where I had previously just been mulling over ideas, I became highly focused. Within a brief period of time, I manifested three significant projects to completion. It was like I have been awakened and propelled forward. I have been able to advance several visions in my life, including major life changes.

I used to get altitude sickness on the third day of being in the mountains. The **Prana** tape has eliminated that. My personal growth is accelerated, including releasing old patterns and embracing the new.

Playing the SWE tapes in my office supports the integrity and balance for the office – the environment in which patients have so much change to integrate in their movement toward health. The tapes support a smooth, gentle, movement towards health on the physical, emotional, mental and spiritual planes.

Albina Savoy, High Point, NC

I'm pleased to say that in the two or so weeks of listening to the tapes, I have had many benefits health-wise. My hemorrhoids are not bothering me, and I am having much better bowel movements. Most of my life I have been prone to constipation. Mentally I am able to look ahead, expect fulfillment, and am laughing again.

For more years than I can remember, every morning upon waking there would be a battle going in my head. There was always some thing or someone to fight with mentally. Further, upon waking I did not feel rested because I would wake up quite often during the night. I may wake up during the night now to go to bathroom, however I do not stay awake like I used to. I go back to a very sound sleep. What a pleasure it is to sleep, after years of not sleeping soundly, and wake up rested. In my head there is music instead of a war. Maybe these sound waves will help me sing!! Something to look forward to. Thank you.

Faith M. Davis, Aurora, CO
Since playing the tapes for myself and around my grandsons, who watch T.V. most of the time, I have noticed some differences. They are calmer. I don't have any pains in my legs anymore after I dance for hours. I think positive thoughts about my ex-husband of 30 years ago instead of blocking any negative thoughts. I sent loving thoughts to my daughter who dislikes me as I have custody of her four sons. I look and feel younger.

Chris Ann Mulreed, CMT/Body Worker
Certified Sound Wave Energy Practitioner, Wilton, CT
Eighteen months ago I had a total hysterectomy. Four months ago, I discontinued all hormonal supplements to try a natural approach. I had been using the SWE frequencies, on a limited basis, for the last two months. I sat down to do my two-month evaluation and check my progress. The hot flashes were gone, the mood swings are non-existent, I am now sleeping through the night, (not an easy task when the hormones are out of sync) and there is no vaginal dryness. I love how the frequencies feel energetically on the body, the sounds are so soothing and relaxing. Clients are amazed when I use my pendulum and I select three tapes to go along with their session. I let them read the "I Am" Intentions for the specific tapes that were selected, and they resonate with the tapes before I even play them. I have been using frequencies for the last 4-5 years as a healing modality in my sessions: harmonic over-tones, toning, tuning forks, Dinshah color, colored crystals, essential oils, flower essences, etc. Nothing has been as user friendly to the client, and as easy for them to incorporate in their daily lives. I just purchased the equipment to play three tapes at a time. I feel we have yet to discover the full extent of the healing benefits of these frequencies. I shall be forever indebted to the client who brought me Nicole's audio tape and said, "What do you think about this?" I think Nicole has given us a gift and I'm committed to sharing it with everyone. My mission is to spread "the word."

Dr. Elaine Ferguson, MD, Chicago, IL

Graduate of Brown University & Duke Medical School

(Author of "Healing, Health and Transformation: New Frontiers in Medicine")

My mother experienced a major stroke in the early part of 1997. I Played the **Circulation** tape for her in her hospital room. . .Her neurologist felt that she may never walk again and would absolutely never regain any use of her hands. Yet, within two and a half weeks after having the stroke, she was walking, and had regained a significant amount of use (50-60%) of her hands.

A couple of months later, Mom returned to the hospital due to an adverse reaction to the medicine she was taking. I used the **Spacelight** tape in her hospital room for a couple of hours while she was in a catatonic state due to taking Haldol. The next day she made such a dramatic improvement that her doctors could not believe it. . .These tapes work!!!

Maud Sejournant, Santa Fe, NM

When I heard my **SoulNote,** which was my first contact with the tapes, I felt a great emotion, like a homecoming. Finally I was hearing myself at a deep, deep level. When I hummed my note, my whole being felt it was expanding, growing as large as the whole universe. It was as if I was seeing my self in a sonic mirror. I was looking at my soul image in sound! The resonance touched every cell of my physical being, as well as my emotional, mental and spiritual bodies. I was the drum, I was the drummer, and I was the tone all at once.

Jac Boyles, Orlando, FL

I started out purchasing the **Chakra #4** tape, the **Cell Rejuvenation** tape and the **Relaxation & Calmness** tape. Then I got the **Chakras #1, #2** and **#3** tapes. After using them all for a couple of weeks, I really noticed the difference in my body.

When I eat now, I never have any problems with gas. I can eat just about anything I want which I could never do before. I've even eaten junk food that I haven't been eating in a long time, and I seem to have no problem digesting it. I have no problem with dairy and no problem with sugar. I have probably gained about twelve pounds of muscle and even lost fat, because my body is metabolizing the food much better. My eliminations are sometimes two or three times a day. My circulation and lymph systems are working much better. So life is just wonderful. It's like Nicole's little booklet says: "Life can become a continuous subtle state of bliss."

God bless you and the wonderful work you are providing. It has only been a week since I began listening to your SWE tapes, but what a week! There has been a very subtle change in my life. But what a difference. I've changed the way I'm thinking, therefore everything has changed. Actually nothing has changed, just the way I react has changed. All the little things that irritated me, the "I don't want to do this!," "I hate this!," and "Why do I have to put up with that"—they're still there, they just don't bother me any more! I don't see them as "chores" or "irritants" now. They just are, so I deal with them, almost lovingly. It's true, I'm in a very subtle state of bliss. I'm not putting off things that I normally would. Things that would build up and bother me until I couldn't ignore them any longer.

Whatever the problem is, I think it through, map out a course of action, and do it! No anxiety! I've also noticed a change in my body. My muscles are stronger and firmer. I've gained muscle mass. I have encountered several situations this week that would normally have sent me into a panic. But I'm able to remove myself from the situation and deal with it in a calm manner. No panic, just results! I'm creating what I want in my life. Everything's flowing! I find myself giving up thanks to God all day long! Once again thank you. Its so nice not letting "fear" control my life. I'm replacing my fear issues with calmness and love. God Bless.

Later, from a phone interview with Jac:
About two weeks ago, I noticed that my seventeen-year-old pet cat, Deedee, wasn't feeling well. I figured she would probably get over it, but she never did; she just kept getting worse and worse.

Monday morning of this past week, I noticed that she had lost a tremendous amount of weight and she was hardly able to walk. So I took her to the vet, who took a bunch of blood tests. Come to find out, she had feline AIDS. Also, according to the blood tests, her kidneys had practically shut down. There were a couple other things he pointed out that were exceptionally serious. He said he wasn't sure if they were due to the kidneys or the feline AIDS. Anyway, the vet told me that because she was seventeen years old, even if we were to treat the kidneys, because of the feline aids she would probably start to have other recurring infections and everything. So he suggested putting her down. I couldn't do it after having her for seventeen years. I brought her home and was preparing myself that she was probably going to die within a day or two, because she had even stopped eating. Then I thought to myself, I beat AIDS, why can't my cat? So I started putting her on the Essiac tea, and I have Nicole's tapes, so I began playing the tapes over and over *where my cat could listen to them.*

The very next morning, after I finished meditating, I noticed that she was sitting on top of one of the speakers. I had three tapes playing, so I was curious which one it was, and it was **Chakra #2**, which is for kidneys. I got some food and water and I set them down next to the tape where she was sitting, and that was the first time she had eaten in I don't know how long. I had been playing the digestion tapes also. That was about two or three weeks ago and she has gained weight; she was practically skeletal, you could see every vertebra in her backbone, and her little head looked like a skull.

I forgot to tell you that I think she had a seizure the day after I brought her home from the vet clinic, because she cried out and fell over, and she couldn't get up or move, so I picked her up. It was after the seizure that I started playing the tapes. I can't believe the change in her, she's jumping up and down off the couch, she's talking to me like she used to in her normal voice. Before, when she would meow it was a very faint, hardly audible meow. Her eyes were sunken in her head, and now she looks normal, except she looks a little thin, but she has still gained quite a bit of weight already and I think she's going to pull through.

I keep playing the tapes, I keep giving her the Essiac tea and I keep saying prayers for her. So I think that's pretty miraculous when the vet said she was near death and her kidneys had shut down, and that she had feline AIDS as well. So there's something pretty miraculous that has taken place here. And I'm very thankful for it because I love this cat, she's a part of my family. Its a miracle for me, its definitely a miracle!

Atoni Dancing Thunder, Sedona, AZ

I have had numerous little healings with the SWE Tapes. It began with earaches, swollen Eustachian tubes, toothaches and headaches that wouldn't go away, even with prescription drugs. Although I have never had a problem with hair loss, my hair is healthier and shinier, and my skin is in better condition, from using the **Hair** Tape.

I have experienced emotional releases that have facilitated an increase in my mental, emotional, and spiritual growth (with a few healing crises along the way). My blood pressure dropped 39 points in two days, from using **Circulation, Cell Rejuvenation, Chakra #4,** and **Compassionate Heart.**

I also had a very interesting experience using **Chakra #2**. I had just soaked in a tub with various herbs, and was lying on the bed with a castor oil pack across my second chakra, (I have had menstrual problems all of my life, and now I'm pre-menopausal). I decided that I should put Chakra #2 on. After only a few minutes, the areas where chakra #1 and #2 are located began to vibrate. It became more and more intense, until the entire bed was vibrating. Needless to say, I was a little frightened at first, but I realized this was just a massive release taking place. Eventually the vibration moved through the first chakra, and then totally released. This was an extremely powerful event, and my menstrual problems have greatly decreased.

But, for me, the entire four series would be worth it for the frequent, vivid and sometimes intense dreams I am experiencing. The information and direction I have received has been extremely valuable.

I am grateful to be able to share these frequencies with others.

Charlotte Sternkind, San Antonio, TX
By 1993, I realized that my fatigue was not going away, I was no longer able to maintain my weight and I was starting to experience fire-ant stinging in my feet, but perceived myself as way too busy to go to doctor. By July of 1994, my body demanded attention. During the following three months, I developed a new symptom every two weeks, which by that time included blurriness of vision and chronic floating patterns, ringing in the ears, loss of short-term memory, dyslexia, numbness or stinging over 90% of my skin, bladder and bowel problems, muscle spasms that left the tissue in knots, and balance problems. While it was comforting to know that I didn't have 47 other diseases, I was left with the big question: What is wrong with me?

The day the neurologist told me, "PROBABLE MS," I was numb. They could not cure me, they did not know how fast it would progress or what my outcome would be. They only knew that statistically, it would get worse. For about a year, I was a good little patient, taking my Tegretal for nerve pain, and Amantadine for fatigue, and slowly getting worse. The last time I was at the neurologist, he offered me Baclofen and a regimen of Betaseron ($846 a month). One day, I took off my shoe and had a rock in it and a huge blister – WHICH I COULDN'T FEEL. That day, I decided it was better to live in pain than be a zombie.

I canceled most of my obligations in my life by using my illness as an excuse. Over the next year and a half, every symptom got a little bit worse. I knew that traditional medicine wasn't helping, I was thoroughly ingrained in the process of victimization. Could it be possible that I could do something about it? I came to realize that mind and body are not separate, but that each of us consist of MIND-BODY, energies rotating together, partly on a physical plane, partly ethereal.

In desperation I asked, "PLEASE, GOD, HELP GET ME THROUGH THIS NIGHT." What I received was, I believe, an angelic healing. I experienced a warm, joyful feeling outside my body.

One day, I opened the weekend news to the wrong page (I was looking for a movie), and there was a tiny ad for an event at one of the local metaphysical bookstores. "Come Hear Nicole LaVoie speak on Sound Wave Energy." My intuition said GO! By this time, I was starting to heed my Higher Self's suggestions.

Nicole was this petite lady with a strong French accent and a fascinating life story. She had these weird sounds on tape. I don't know if anyone else felt what I felt this day. The **Circulation** tape made my body warm and my hands hot. When she started playing and talking about the **Love** tape and how it sometimes provoked anger in people who weren't open, I felt a strong energetic pull from the middle of my upper back out through the top of the head! Based on these two feelings, I purchased and took home my **Basic Series** and STARTED CHANGING MY LIFE.

My husband thought I was crazy, buying noise on tapes. The cloud had a very heavy silver lining. **Chakra #5** loosened my tongue, and I was telling the TRUTH for the first time in 18 years – and because of **Chakra #6**, he was listening. We started talking about issues I'd written about, and a great many changes have come about in our lives.

I used to be a subservient washrag with a workaholic husband; now, the **Chakra #3** tape has shown me that I AM a powerful being. I'm also taking art classes just for the fun of it. I am much less critical of the outcome, and spend less time in thought on WHAT I'm going to make. I'm learning to let my creativity flow.

I enjoy the heightened dream state from the MEMORY tape. I often get good information on transitions I need to make, as well as being able to remember them long enough to write them down in my dream journal, which I later use for analyses as well as drawing or writing poetry from the dream.

A great combination is the **Clarity & Focus** and the **Public Speaking** tapes. I use the former for every paper I write for my college classes, and the latter when I have to do an oral presentation. Even with my Toastmasters experience, the two tapes allowed me to give a whopper of an inspired eulogy that I delivered at my father's funeral about my angelic rescue and to have faith in higher powers. I also play the **Communication** tape during my presentations. Every talk I give now becomes an awareness of my ability to be both teacher and student.

With **Spacelight**, I have cleared my own house, and cleared one other house of an entity. As far as I could determine, this spirit had no evil intent – I certainly was not aware of its energy before the tape. But it decided it didn't want to leave. The first three attempts in playing the tape, I left the room. The fourth time I sat with the player to finish playing the tape all the way through. The first time, the entity threw the bathroom trash back out of the wastebasket. The second playing, it stopped the tape in the middle; the third playing, the cord was pulled from the back of the player in the middle of the tape. To the best of my knowledge, the entity finally left for "lighter" climates.

I used to take about $200.00 worth of vitamins a month just to stay awake. Now I no longer need them, and the only thing I still am directed by my guidance to take, is some Omega-3 which runs me about $21/month. As you can see, from a purely financial aspect, I've already made back my investment in the tapes.

I like the **Merkaba** for meditation and the **Energy, DNA, Enzymes** for the few times that I really become fatigued, especially after healing sessions.

Lately, I'm directed to the **Purification** frequencies a lot. I've been going through some healing crises from both frequencies, and as I learn techniques for improving my healing skills. By the way, I was able to start doing hands-on healing about 6 months after I got all the frequencies (I attribute this to my rise in consciousness from the tapes), and for about 4 months before I finally got my Reiki I attunement. I am already working on the emotional level which my friends tell me is primarily a Reiki II capacity. On May 1, 1998, I was ordained as a minister. It seems to me that I am already making positive steps towards my new profession.

Things have been changing a lot in my life. Thank God, Nicole, and Sound Wave Energy for the frequencies of **Transition** and **Gateway**. They are helping me assimilate all these changes with grace.

Recently I was at a lecture given by an animal communicator in

San Antonio. She ended her lecture by saying that her animal friends had given her some advice for us. Her own cat's comments were, "Purr more often. You will find balance through doing this." It occured to me that that is exactly what the **SoulNote** is—it is in effect the human equivalent of the purr.

The day that I had my SoulNote defined by Nicole, I had a major healing crisis on the way home. Suddenly, I started crying—out of the left eye ONLY! This lasted for more than half an hour, perhaps my feminine side was finally releasing all the pain I had caused my emotional body by attempting to deny the woman I AM her right to be.

After I had the SoulNote done and started humming it before saying the intentions for the frequencies, I noticed a dramatic increase in the effects of the tapes. Also, whenever I use my pendulum, if my intuitive nature is less than 100%, I can "bump" it up by singing my note before asking any questions. I use it to manage pain in my body whenever I'm not carrying the **Emotional Release** tape, and I'm happy to say that the intensity of the pain has dramatically subsided over the one year that I've been using the frequencies.

Christine Shelton, Reston, VA

For the past ten years, I have experienced long term pain throughout my body. The doctors could not determine the pain's origin. I had a fall ten years before the onset of the pain, and the doctors surmised the fall may have been the problem. I ended up with a barrage of varying symptoms that played constant havoc throughout my entire body. Any small physical task became overwhelming for me to do. I would need to rest after cooking a simple meal. Standing and sitting were difficult. Driving a car for a short period of time was painful. I had great difficulty sleeping at night. I could not feel my bed, all I felt was the pain. The pain and discomfort took on a life of it's own. During this time I made many visits to allopathic doctors, chiropractors, physical therapists, massage therapists, acupuncturists, homeopaths, nutritionists, energy and spiritual workers, and the list goes on. I would experience varying degrees of relief from all of the different healing modalities that I pursued. Sometimes the pain would be diminished for weeks or months, only to reappear in another form. I was extremely frustrated. The physical pain was also taking a toll on my emotions and spirit.

My search to heal led me to take many holistic classes. At one of these classes I had the good fortune to meet a wonderful holistic consultant. Together, we worked on various mind, body and spirit modalities to get to the "root cause." At a spiritual and emotional level I was having good results, yet the pain in my body persisted. Several months went by, and I felt I just was not getting it! I felt I was "on the path," doing all the "right stuff," and I was still in physical pain. I called the consultant and shared my frustration. I said I was tired of the tears, tired of the fear of not knowing what was wrong, and this was not the way I wanted to live my life. This was not living!

The consultant told me about this incredible lady, Nicole LaVoie, who had created frequency tapes which helped to facilitate wellness. I personally did not know of anyone using frequency tapes to heal, but I was ready. I ordered the tapes and played them as much as possible. Within two to three weeks I knew I was experiencing positive changes. When I would talk to my family on the phone they could "hear" the change in my voice. They asked me what had happened? They could "feel" the energy in my voice. I told them I was listening to tapes with frequencies on them. They did not really understand. All they knew was that I sounded wonderful. So they said, "Well, just keep playing the tapes"!

Because I was feeling better emotionally and physically, I was able to take a look at deeper issues regarding my circumstances. As I did this, answers came to me about what was the "root cause" of my pain. As these opportunities continued to present themselves everything became less fragmented, my energy continued to increase, emotions soared, and I felt good. One night, just before falling asleep, I started crying. I was crying because I knew I was going to be all right. They were tears of sheer joy. I **knew** I had passed through the dark night of the soul. I **knew** all was well.

I now find the comfort of my bed, when I lay down to sleep (as opposed to pain). My daily life takes on wonderful new aspects of healing. I prayed for "Joy"... now, I **know** Joy! I am a testimonial to the SWE tapes' ability to facilitate healing. I am grateful for the opportunity to have experienced these incredible and beautiful tapes.

Carol Osburn, Seattle, WA

Profound... is the only way to describe the effect the SWE tapes have had on our lives.

My goal and quest (as I stated to a friend several months before I first heard of SWE) was to raise my vibration, and to stabilize and capture the long-term, good benefits we were getting from our meditations. The minute I heard about the SWE Tapes, I knew they were exactly what I had expressed a desire for, and I ordered the **Basic Series** the very next morning.

Our tapes arrived just before we were scheduled to take a short trip, necessitating putting our cat in a boarding kennel thirty-some miles from where we live. In the past it had always been a traumatic trip for all of us – cat would be yowling in terror, over and above our soothing reassurances that she was going to be OK. As it turned out, we were playing the **Chakra #1** tape in the car, and this trip was decidedly different. As we drove, she settled right down – calm and relaxed – on my hubby's lap, and was only mildly concerned about transferring to her new quarters. What a contrast!

As we continued on our way, my husband drove while I read "Return to Harmony," with the tapes intermittently playing in the background. When we arrived at our destination, my husband's low back pain had greatly decreased; not at all the usual result of a five-hour car trip. Instead of being tired, both of us felt refreshed.

Over the weeks that followed, both of us seemed to need less sleep, have more control of our thoughts with better ability to focus on tasks, we were more relaxed and centered, and most of all, we had a feeling of comfort, security, and a decided sense of well-being.

However, on the job scene, all was not well. I was going through a particularly troubling time, not coping well and feeling powerless. I was easily intimidated and terrorized by aggressive and sometimes rude telephone customers (in whom I seemed to bring out the worst), and mirroring their behavior was not a good plan. I felt I was being "victimized" on the phone, and at work as well, but I could rarely speak up for myself and get my point across; this was an extremely frustrating situation.

Gradually, over the months of devotedly listening to all of the series of SWE Tapes, I began to change. How do I describe it? I felt better about myself and more centered and grounded. The impatience, irritation, and anger, that had been so evident before in the difficult times, was no longer a part of me.

I could speak my truth, people heard me, and even if they didn't agree, they were not offended.

Although I still spoke all week long, with hundreds of customers with problems, I didn't take them on. I just dealt with them, naturally and compassionately. It seems as though they had lost their knack for "getting to me." I could empathize, but not get caught up and carried away by their negative emotions, as I had in the past. I was in my strength, inside myself.

For many years I had carried a festering anger and rage (sometimes thinly veiled), at being a physically abused and unloved child. Now after years of trying, more or less unsuccessfully, to overcome that emotional pain, I can say it is no longer a part of my experience. A lot of anxiety and resistance are gone as well.

Thank you for allowing me to share with you a chapter of my healing and transformation.

P.S. The cat is considerably more frisky and playful now than when we adopted her four years ago.

Holly Shrauger, Orlando, FL (from a phone interview)
My name is Holly and I work with *Three in One Concepts*, using *One Brain* techniques. About a month ago I had my **SoulNote** done with Nicole. I was very fascinated and impressed with her as a person, with her background and with what she has put together with the SWE tapes. I purchased a complete set and began using them right away for myself and my clients. I work with one client at

a time, and I select the tape that seems appropriate for that session.

I feel that we are all energy and vibration and that the goal in life is to raise our vibration. We can do it gently and in a very non-invasive way with this work.

When I found Nicole and the SWE tapes, I knew that this was another piece of my work with people. The tapes are another level of "doing it gently." When visiting with her, I found that the **Mental Series** had to do with integrating right and left brain and overcoming learning disabilities. I was very excited.

When I am working with a child or an adult with learning disabilities, I play the **Mental Series**; **Brain, Courage & Prosperity** and **Clarity & Focus** tapes. They are excellent. My feeling is that I can be extremely accurate determining what each person needs, because of the muscle testing. Their body will tell which tape and how many tapes to play during a session. I play up to three tapes at a time. I have found it to be wonderful. Quite often I will muscle test for a tape and it turns out that the person needs one that I would not have chosen. But the tapes always match the issue that they are working on at the time, so I feel that it enhances my work.

One of the first experiences that I had was with a gentleman who wanted to work on his reproductive chakra. I played the **Chakra #2** tape. At the end of the session, he could feel that the area was open and felt very good. He was very pleased so he came back the next week. In his background, he revealed that he had been badly abused since he was two years old. A lot of different issues were going on as well, he had lost a lot of self-confidence and mental clarity, because of these beatings. His learning disabilities are very severe; in fact, that is why he was sent to me. He had been working with a gentleman who suggested he come for the *One Brain* work. Now, with the tapes as well, I feel like he's making major breakthroughs. So, that's exciting! I was really pleased.

By that time I had been listening to the tapes for a couple of days, and nights too. I find that my clarity and creativity during the sessions have been tremendous, I feel much different personally. Of course I'm also listening to the tapes while I'm working with clients. I'm much stronger and my physical discomforts are going away. I know that some tapes are working to balance my vitamins, minerals and hormones. I am experiencing a little discomfort from listening to the tapes from time to time, but it just moves on through.

One little boy, whose grandparents said "put him in an institution," came to me. After the first session, in which I played the Mental Series tapes, it was clear that he had a severe learning disability. He is in the sixth grade, should be in seventh, yet can't read basic words. He can only read words that he has memorized. He is very much a behavioral problem. He went home after one session and nobody—including the parents and grandparents—could believe that he was the same kid. So, I feel that the work I do is being enhanced by the tapes. I think the two go very much hand in hand. It's wonderful.

Right now I'm teaching a *One Brain* class. So, I'm playing the Mental Series constantly all day; everybody is calm and the students are getting the work done. All three in my class have learning disabilities. Normally, because the students are learning new material, and because their own things are coming up, they are really stressed out. With the tapes, they are all calm at the end of the day. They all feel that they understand the material that has been presented. I return home and I have energy instead of being tired. It's great!

Yesterday, I had a couple of migraines. I'll be listening to a tape, then all of sudden, the pain breaks. I just feel wonderful. So, I think my blockages are opening up. One of the things you said about the tapes is that they can balance the hormones. I think my hormones are changing because I'm in menopause, and I have been real pleased about that because I didn't want to take estrogen.

I'm very pleased to have found the SWE tapes and to have found Nicole. I feel that she is a very sincere and aware person to have put this together. It's just a real gift and I plan to continue to work with it for myself and my clients—and to learn more about it.

Lynn Veitch, Winter Park, FL
I was introduced to Nicole LaVoie and the SWE tapes about one year ago. I knew immediately this was something extraordinary and wanted all the tapes. I began listening to **Chakra #1** and **#2** all

---

night. After only two weeks, I realized I was not experiencing the usual PMS symptoms I usually did each month. No aches, cramps, heaviness or bloating. This has continued ever since; what a relief! In the meantime, my husband who was also exposed to the frequencies, began to experience relief from allergies that had kept him awake at night for years.

We both began to notice an opening of creativity in our lives and took steps to demonstrate that in our lives, whereas before we held back more. I have also used the tapes in sessions with clients, I use breath work and body/mind therapies. I found that all report a heightened sense of awareness and a deeper connection with themselves. This has ranged from profound spiritual experiences, to feeling very grounded in the physical body. Occasionally, I am fortunate enough to assist in the birthing process. Recently I played the **Relaxation & Calmness** tape for a young woman who was in labor with her first child. It seemed to assist in supporting the birth, which was one of the smoothest and fastest I have ever attended.

Finally one of the most amazing results that stands out for me is the healing of my twelve-year-old dog, who became very ill. When the vet could give no diagnosis or treatment, I tried chiropractic adjustment and the **Cell Rejuvenation** tape. Within several hours he was jumping around like a puppy again.

I continue to see endless possibilities for the tapes and am excited that such a wonderful tool now exists to accelerate our inner journey.

Sandi Christopher, Stewart, FL (from a taped phone interview)

My mouth blew up with this abscess. It was unbelievable. I looked like I had a baseball in my mouth. The infection made three pockets, including one underneath my neck. We did everything under the sun, from healings, to Reiki, to herbs, to aloe—and nothing was helping. I even went onto tetracycline and all that was doing was making me sicker. So I played the **Bones** tape and the **Cell Rejuvenation** tape an hour here and an hour there, just as I had understood you to recommend. I stuck with that.

Finally, one day, I had stopped the tetracycline and everything else, and I was getting really sick. I wasn't even able to get up off the bed. I got very angry and said I was going to listen to the tapes more than the hour at a time you had instructed. Every time I would listen to the tapes, that pocket of poison would vibrate, so I knew something was really happening.

After six days in bed, I was tired and mad at myself for being sick. I had stopped everything except the tapes. After listening to the tapes for an hour and a half one time there was a big explosion in my mouth and everything drained out. It was absolutely wonderful.

So, that's my playing with the tapes so far. Except a couple of other things, like playing the **Love** tape subtly, almost inaudibly, and watching the kids and animals mellow right out. Also, since I've been playing the third and sixth chakra tapes together for a while, I lost about six inches around my waist. I feel the vibration in my fat when I play the tapes. They just jiggle everything inside of me. I feel like there's Jell-O moving around in my body.

A lot of old anger that was unbelievable, and that I didn't know was there, has been surfacing really easily and moving through quickly. I actually feel that the abscess was part of that anger coming up, anger that I had been unable to get out. So, it's working, and thank you so much. We love you here.

Bill and Lisa Cucuel, Winter Springs, FL (from an interview)
We started with the **Chakra Series** of tapes, beginning with the **Chakra #1** tape. Bill and I and the little granddaughter played it several times, on great big speakers, throughout the first evening. We could all *feel* the tapes. Interestingly, I could feel the Chakra #1 tape in my head. The next day my daughter had diarrhea—no, more than diarrhea, a complete elimination, twice as much as usual, and old stuff.

The next day we started to play the tapes for **Chakras #2** and **#5**, all day. By the evening, my senses, especially hearing, were so heightened and sharpened, almost like a drug, except it was clear, not foggy. It was a little disturbing because it was so different, not anxious, just disturbing, until it mellowed out.

On the third day, I was doing my meditation with the tapes on in the background and it was totally different. It was not unpleasant. Normally, I have light experiences. While playing the tapes, it was a vibrational experience. That may not sound like much but it was very exciting to experience. It felt like it was faster—not hurried, just a quickening.

I noticed, after the third night since beginning to play the tapes, that I slept deeply all night long for the next three nights. My prior restlessness was from neck and shoulder pain that I cannot remember being without. All my tension goes right there. I realized that the good sleep was because my neck and shoulder pain was gone, and now I can turn my head normally again too.

Then we started playing the **Love** tape at night. I woke up that first night with what felt like indigestion, which I never have. It felt like there was a painful lump in my upper chest. I finally guessed it was the new thymus opening up, as you said would be happening for all of us. Three more times I was awakened with this same ache in the upper chest. In the morning I felt wide open right here—right in the thymus area. It was very refreshing and beautiful.

I thought at first that the **Relaxation & Calmness** tape should be called relaxation and tension. I felt tense when it was on and then I relaxed when it shut off. I guess that I wasn't used to the true relaxation vibration yet, I didn't recognize it. After the first couple times, it was relaxing and calming to hear and feel the tape playing. Now I look forward to it. I feel more accepting now and find that the old relationship hooks are not catching me up any more.

I've been having problems with my left ankle for eight months. I tore ligaments and really damaged it. After just the second day of listening to the tapes, the swelling has reduced noticeably. I can feel where the tears of the ligaments are and feel that they're filling in. I was amazed at the reduction of the swelling. I'd had a whole series of acupuncture for the ankle and other things, like chiropractic adjustments and prescriptive anti-inflammatories, and none of them helped, until the tapes. This is quite remarkable to me. [Note: In a subsequent conversation, it was revealed that the ankle pain was completely gone after ten days.]

We listened to the **Christ Consciousness** tape all night last night. You told me not to play it too much at first unless I wanted discomfort. Some stuff is coming up and you were right. I woke up feeling feisty, a little short tempered. It'll be interesting to see if anything else comes up today. This is powerful stuff.

We've been writing a book. We were into the first rewrite as we began to play the tapes. The rewrite has been going extremely well. I've noticed that I have a lot more clarity and energy for the project. Things in the book that I thought were finished, I've been finding big errors in while playing the **Clarity & Focus** tape.

Since using the tapes, I've felt a clearing of some old resentments that I was totally unaware of, with people in my family and in my circle of acquaintances. In fact, it's funny—two of the people who came into my awareness as individuals that I had resentment for and then worked through, immediately called on the phone, so I got a chance to completely clean up those old resentments. That poison just couldn't stay inside me. The clearing wasn't violent, it was smooth.

I've gotten guidance to save the **#21** tape for my birthday next week. I will go easy with it as you recommended.

Deborah Lynn Darling, Garrettsville, OH

I have never been very good at following orders, nor have I ever considered myself to have many fears. I would have to say that I always thought I was pretty fearless and I think almost everyone who knows me would have agreed—that is, until the **Christ Consciousness** tape. This tape was one of the first tapes that I bought from you and though you told me to take it slowly, I decided to listen to the tape all night. All at once I went from being fearless to paranoid. Talk about bringing up fears I never would have believed I had. I was afraid to leave my home, afraid someone would steal my car etc., etc., etc. My friend and partner Jill Lawrence said "it's the tapes," and I of course said "no way"—but it was. I stopped listening for a few days and then, when things started to go back to normal, I began to listen to the **Christ Consciousness** tape as recommended. This tape now brings me great comfort and has helped me to remove all of those hidden fears.

You would think that I would have learned my lesson by the

time I got to **#21**, but once again I tried to listen to the entire tape. The Universe gave me a quick message however, it made me sick to my stomach and dizzy. However, the very same tape that made me violently ill is now one of the tapes that brings me the most calmness today. Thank you so much for your work.

Dr. Alexander Fries-Tersch, Vienna, Austria

I am very glad to send you my report about my experiences with your SWE tapes.

It is now two months that I have been using the tapes and from the beginning I was using them quite intensively, i.e. two to six tapes per day. My first tape was **Chakra #1** and I very soon felt this chakra to become warm and active. Three days later the circulation in my right leg was nearly normal again—I used to have ankle swellings.

Three weeks later my digestion got very much better, elimination was much more complete in quantity and in quality. More waste products went out. Breathing became easier although the nasal septum is still incurvated. The nasal passages seemed to open up more.

Having had a heart condition some years ago I had to be very careful with exertions. Now I find that I can exercise much more easily than even as a young man.

The cleansing effect of the tapes extended even to the teeth, two of which started to hurt and had to be root canal treated. There are other changes too, my body feels different but I can't put a name to it.

Emotionally there is a lot of mostly unconscious clearing and this process is still going on. However I am now able to listen to the **Christ Consciousness** tape without reactions which used to be very strong. The same goes for the **#21** tape.

Mentally, something quite unexpected and interesting happened—my thought processes slowed down considerably. Instead of buzzing in my head like a cloud of flies, they are walking sedately. And a recent brain field analysis showed a marked predominance of the right brain. My impression is that this means a sort of balancing out as I have all my life certainly been a very left-brained person. Where it will end I have no idea as the process is in full swing.

The more I use the tapes the more I feel a tingling in the layers next to my skin. It feels like a good blood circulation and helps that certainly but I think it is more an improved energy circulation.

I find that I can deal with people more easily because I am less defensive. You told me some weeks ago that my heart was more open and I feel more friendliness and sympathy meeting all sorts of people.

Regarding my hearing, which has been somewhat impaired for a couple of years, at first there seemed to be no improvement at all. Now my impression is that the hearing ability is stabilized, meaning that the hearing level seems to remain the same instead of being irregular.

Although I am listening to the **Spiritual** and **Ascension Series** regularly I can't say anything about spiritual changes. Maybe I am more grounded and sometimes I just know without having it all reasoned out. My guess is that there is still a lot of clearing and cleansing to be done with the chakra tapes before changes in the higher energy levels can take place or can be noticed.

There is a lot more to tell. For instance my dowsing ability has changed and I am less sure of it on the one hand and on the other I do observe new patterns of the pendulum swing. My cat, Hercules, is behaving differently and gets mauled by dogs and scratched by cats. A lovely friend on four paws, he seems to be less able to protect himself.

With the soul note there were no obvious changes but a dowsing friend told me that it gave off a strong and harmonious field.

This report is as comprehensive as I could make it but there is surely more to come and I am looking forward to it. Thank you very much for these wonderful tapes and your advice.

Bones

Jill Lawrence, Canton, OH

Ever since I injured my knees cheerleading in high school, they have given me problems from time to time. Three and one half decades later, the knee situation had truly been exacerbated. I was having so much trouble that, at age 52, I had to purchase a cane to get me through the most severe flare-ups. One or both knees would simply crumple out from under me without warning. In addition, on occasion, I would suffer knee pain significant enough so that when I walked, the pain was so sharp that I would involuntarily cry out.

I can truly say that this is in the past now, thanks to the SWE tapes. I say this without exaggeration. Using SWE tapes is the only thing that I did differently. Within a month of playing the tapes, virtually all knee problems vanished! Totally and wholly! No more crumpling and collapsing under me and no more pain! And certainly no more need for a cane!

Marion Light Ray, Silver City, NM

Last summer when I was running, I tripped and fell, banging and injuring my knee badly. I was so embarrassed, falling in a public place, that I stood right up and walked away briskly. My friend said to me, "It is a good thing you didn't break something. You know how it is when you get to be seventy, bone breaks don't heal as easily." I thought I'd recover full use of my left leg and knee, yet the flexibility in the knee failed to improve. I could no longer sit on the floor with both legs tucked under me. I could not do any daily exercises that included knee flexibility.

When I returned from a cruise with my SWE tapes, I began a program of recovery for my left leg and knee. I played **Chakra #1** with the **Bones** tape as I did my daily exercises. I soon began to notice an increase in my knee flexibility. Now, one month later, the knee is healed. I can once again sit on the floor with *both* knees tucked under me. I wonder how many other miracles can happen as a result of playing the SWE tapes as a background to daily activities.

When I had played side one of **Chakra #5**, my miniature dachshund Gaia began coughing, so I took her outside. My neighbor invited us next door for a visit and my Gaia went with me as she loves the treats my neighbor usually has for her. Inside, Gaia began coughing again but now she began spitting up a clear liquid as well. So we returned home where I played side two of the Chakra #5 tape. The coughing and spitting continued and I noticed she was now shivering badly. I turned the Chakra #5 tape off and just held Gaia in my lap, stroking her. As I stroked her, I wondered if the Chakra #5 was just too much for this little dog. Then I heard the inner words, "What turns it on will turn it off." Right!

I played the Chakra #5 tape for the remainder of the day. Next day the shivering had disappeared and only an occasional cough from Gaia, but no more spitting up. Later that morning my neighbor phoned to tell me I had better get Gaia to the vet. Two other dogs in our neighborhood had come down with the same symptoms and their owners had rushed them to the vet. The dogs were diagnosed with pneumonia and given shots. I assured my neighbor that Gaia was fine. Just to make sure, I once again played the Chakra #5 tape together with the **Circulation** tape.

D. Johnson, ME

Our son, J.P., was born with an absent fibula on his right leg. Since his birth we have consulted with some orthopedic surgeons who have put the leg in a cast and tried allopathic treatments of just keeping the leg stabilized. It was suggested we use Nicole LaVoie's SWE tapes: **Circulation, Bones, Chakra #1** and **Relaxation**.

We have been using them daily for about a month. Then we had an MRI done and there was a half inch of growth in the fibula, extending from the heel to the ankle! That is very encouraging, because it opens up an entirely new type of treatment which is available to us. We are very pleased with that and so is our new doctor who is a member of the National Institute of Health.

He says there is a possibility of getting the cartilage to turn to bone: he was mentioning something about a "good benefactor," made by some Biotech companies. They have a product that is approved. It sounds similar to the shark cartilage therapy. But it wouldn't have been possible up to this point *without this initial growth*. What he also said is that in addition to the half inch that has solidified, he felt that there was cartilage through half of the leg, at least. In other words, the cells have turned into cartilage, and it is now a matter of whether it solidifies into bone. That is the next test.

Cora Tremblay, Miami, FL

I received my **SoulNote** tape. I love it and thank you much. Let me tell you my reaction. Surprise, because I remember a long time ago when I was young I used to hum this note but this one is more powerful. The higher frequency on the other side made me jump but I was able to listen and my right shoulder became painful. The next day it was my left shoulder up to my jawline and after that I became very relaxed like I never was before. Even after the sound finished I stayed relaxed without moving a muscle for a while and it made me feel good.

Also last week after listening to **Christ Consciousness** and **Chakra #3**, I heard a choir of Angels singing. I never heard this song before, I stood up and it seemed to be from the outside and I looked outside but there was nothing. It lasted for maybe five or ten minutes and it gave me such a beautiful feeling that I hope this song comes back again.

Barbara Maynard, Santa Fe, NM

The **Basic Set** of tapes have proved to me their "greatness" by healing my physical problem by 90% release of burning, pain and cramps due to nerves being pinched between my third and fourth lumbar. The excruciating pain for several years has for the most part left and it is "Divine" to be without pain.

Also, I just went through a home move, doing the bulk of packing and moving myself, leaving the large, heavy things for the movers. I lifted and carried boxes etc. that in the past would have been impossible without throwing out my back and atlas and axis vertebrae, causing extreme headaches. For a month I worked at this moving without a headache. . . Hard to believe! There's another miracle!

Cervix

Cheryl Kerr, Orlando, FL

I am so happy I met you. I appreciate the discovery you have made with the SWE tapes. In January of this year, I went to the midwife who has been examining me since 1992. At that time there was a growth on my cervix that was red, irritated and bleeding. A Pap Smear showed some abnormal cells also.

Today, there was a great improvement! Usually I am so tense and anxious I can't be touched without a lot of struggle. I was very relaxed and the growth has improved: it does not bleed when I am examined! The midwife was very pleased and so am I . She said it looked as though the tissue was part of the surrounding area, more normal in every way! I played the SWE tapes you recommended, as well as taking some very good products.

Over the last few weeks, I've been really excited about the tapes that I received from you. I've listened to all the **Chakra Series** and the **Christ Consciousness** tape. My most exciting experience came through the **Love** tape. What happened was, after listening to this tape one night, the next day I received a phone call from a very dear friend who lives several thousand miles away. I have not spoken with him for quite some time, but I have been sending love to him. This person called me and told me how much he loved and cared about me, and how wonderful I was. That was a dream come true, because for so long I have been trying to convey that to him. But I was always a little too shy to even speak those words. I felt so energized and happy that he realized on a very deep level what our true friendship was and realized what our true heartfelt feelings were really all about. That was the most exciting and wonderful experience for me.

Eyes

Grace Torres, Boulder, CO

A year ago in May of 1997, I attended your lecture in Virginia and I purchased the basic series of SWE tapes. At that point in time, I had an eye ailment, blephiritis, that had been with me for 1-1/2 years.

Although my eye condition was not life threatening, it was a personal nuisance and irritating. I tried both conventional methods (antibiotics) and alternative treatments (acupuncture) but nothing completely cured my ailment. After using the basic series for 1-1/2 months, my eyes appeared healthy in a very natural way. I was finally able to wear contacts after almost 1-1/2 years of wearing glasses! I was ecstatic when the tapes finally helped me heal my eyes, among other things. It has been one year since I've purchased the **Basic** set and I am very satisfied with the results, so much so that I own all four series and I am an SWE distributor who firmly believes in this product.

Thank you for making miracles a reality!!

Heart

Nancy Mayes, Acworth, GA

I am writing to tell you of the profound effects that your SWE tapes had on my father the last two months of his life. He had built an experimental airplane and loved to fly. Because of his heart condition, he did not have a pilot's license, but would take the airplane up anyway on occasion. He was flying and ran out of gas. While trying to switch tanks, he crashed the plane. The accident nearly severed his right foot.

While he was in the hospital, I brought my tape recorder and several of your tapes to his room in Intensive Care. I could see a tremendous struggle in his body, there was so much resistance and fear. I had always seen this resistance operating in him. Whenever I had tried to talk with him about healing and the mind-body connection, he would turn his body away from me. If I started to speak to him, or if I tried to put my hands on him to give him a Reiki treatment, he also would turn away. He came close to dying.

I bought the Benedictine Monks' Chant tape and, as often as I could, began playing this with the **Relaxation & Calmness** tape. It took about one week for his condition to improve. A lot of people prayed for him during that time. It was wonderful to see the resistance and fear leave his body. Peace and calm were present within him, maybe for the first time in his life. He loved listening to the tapes and appreciated receiving Reiki treatments as often as I could be with him.

He spent the next two months in the hospital. The doctors eventually released him to come home. The inactivity and the two surgeries put too much stress on his body, even though his foot was healing beautifully. Within the first week he was home, he died in his sleep, listening to the **Cell Rejuvenation** tape and *Enya,* playing as loud as possible on the tape player by his bed. His passing was beautiful and elegant. During that week, when he awoke from sleep, he told my mother that he was having the most wonderful dreams. He thought he had already died and was in heaven. I know he was being taken over to that place in the most gentle manner during this time.

I wish that my father could have stayed. Right up until his death I thought he would rally, as he had so many times before. I know that your tapes made a tremendous difference in the quality of those two months and in his transition.

Thank you, Nicole, for the SWE tapes. You gave my father and my family a tremendous gift through the use of this technology. I am very grateful that this is available on the planet today to assist with our healing and transformation.

Nose and Throat

Rev. W. James Drummond, West Palm Beach, FL

For several years I had been having extreme discomfort in my throat area. I felt like I had swallowed a softball—and while I might have a big mouth at times, this ball felt like it was lodged into the side of my neck. I would have trouble swallowing, and at times even trouble eating and breathing. Sometimes I would experience nausea. When it got really painful, I would inquire with the medical profession and although they ran tests and did scopes, x-rays and the whole bit, they always had a different opinion on what it was and how to treat it. So their remedies resolved the pain for a few days and sometimes a few weeks—but sooner or later the pain would show up again.

Upon our first meeting in your motor home, I asked you to play the **Chakra #5** tape. Having always been interested in sound and light, I certainly was not ready for what occurred that evening. I immediately felt the lump begin to dissolve—and fast. It was a physical feeling. The only description I can think of is like dropping Alka-Seltzer in a glass of water and watching it turn into bubbles. After about five minutes the lump was reduced to half the original size, and by the end of the tape the discomfort I had for years was totally gone—and has not appeared since. At this point I thought that these tapes might be a panacea not just for the physical body but the emotional and mental bodies as well, so I wanted to work with them more.

At this point I think it's worth mentioning that my economic state was not in the best condition. As a matter of fact, I was down to my last hundred dollars. I was not employed and didn't know where I was going to get money to pay rent for the next month. The couple of hours I spent talking with you and listening to the tapes made me believe that it was important to buy not just one, but a couple, and take them home. I purchased the Chakra #5 tape to see if I could learn to speak my truth—and to make sure the pain stayed away. I also purchased the **Chakra #6** tape in hopes of opening my third eye more so that I could clearly see the path Spirit intended for me to follow.

Out the door I went, with less than thirty dollars left to my name. I am telling you this, so if you have the opportunity to share this with others, you can express to them how very important these tapes seemed to me.

That was in February. By the middle of March, your records should show that I had acquired all the **Chakra** tapes and several tapes from the **Spiritual Series** and the **Mental Series**. I understand now that the cost of them is not expensive. I have experienced the greatest return on my money with this investment than from any investment I have ever made.

Recently, when I found out that I could simultaneously listen to up to three tapes, I have been playing the frequencies every chance I get. The results are getting better and better.

There are several other small notes I want to add. Since age sixteen, I've had a broken nose and a deviated septum. It was starting to impair my breathing, but I didn't want surgery. Just the other day I noticed that my breathing has improved. By looking up my nose with a mirror, it appears that the cellular structure could be changing and my nose could be straightening. Also my personal massage therapist commented about the fact that there appeared to be a swelling in the crown of my head. After a good laugh about the swelling-head phenomenon, I asked if she thought it could be that my skull bones were restructuring. She said, "Well I don't know, but your head has definitely changed. I know what your head felt like last time and I know what your head is supposed to feel like – and it's different."

In general, I am feeling full of vitality and energy. I have a much stronger sense of focus, I'm more relaxed and just all around feeling great, which I contribute mostly to the use of your SWE tapes.

My upbringing was to follow the Olden Golden Rule: Do unto others as you would have them do unto you. I truly feel that in our hearts we are all here to find the Highest Good in ourselves as well as others. I feel that we need to honor and care for this magnificent temple we call the body. I give you full permission to share this testimony with others if you feel it will benefit their growth to the level your tapes have benefited my growth. I fully believe that your heart's intent is aligned with the highest Light and Love of Spirit for Mankind. I feel that your work is absolutely necessary for the uniting of the hearts of people and the healing of the planet at this most important time. I honor you for following the path of your spiritual mission. Any healing on oneself is a manifestation in the total healing of All.

Skin

Justine Metteer (Age 13), Pagosa Springs, CO

I got badly sunburned and returned home that evening in real pain. After applying the usual "stuff" I began playing the **Cell Rejuvenation** tape through much of the evening and the following day. The pain was much less by the next morning, and by that evening I had already begun to peel, so I played the tape throughout the day for the first two days. Usually if I get a sunburn this bad, it takes four days before I get to the peeling stage. Now, four days later, it is all better.

Teeth and Gums

Richard Choy, New York, NY

I had a throbbing pain in my teeth that was so bad I could hardly sleep. Eating ceased to be a pleasure. I made an emergency dental appointment. As I sat in my dentist's chair, he reminded me that on my last visit, he had suggested a root canal for that very same tooth. Instead, I opted for a deep filling. Now, he said, I had to face the inevitable.

I told my dentist that I was committed to holistic measures, and that I needed a little time to try some things out. The next day, on reading "Return to Harmony," I decided to try the **Bones** and **Cell Rejuvenation** tapes. Within four hours, the painful throbbing cleared up. Some pain returned the next morning. I then continued to play the tapes for two hours each night and within two weeks the pain totally disappeared even when I wasn't playing the tapes. In addition, normally every one or two weeks, another part of my mouth would swell up and I would regularly get cankers. All this has totally cleared up. I continue with these tapes as a preventive measure. All my teeth feel fine.

Mignon Lawless, Ph. D., Norwalk, CT

I have been using the Sound Wave Energy frequencies for ten months. I am 55 years old and did not take hormone replacement therapy during or after menopause. Since using the SWE tapes, my body has begun to produce estrogen again. My gums were starting to deteriorate and every night for over a year, my gums would bleed in one particular section after flossing. The bleeding has stopped and after cleaning my teeth last month, the hygienist informed me that it was a pleasure to see such healthy gums!

I was taking buprofin regularly for join pain. It just dawned on me that I have not taken any in about 4 months. Also, I don't take

---

supplements anymore, the tapes provide them. I had been taking expensive herbs for elimination 2-3 times a day. Thanks to **Chakra #1**, I no longer need those, either. I have also noticed that I generally sleep one hour less each night. Until now, an unattainable goal was to sleep less. I have thin, fine hair. My hair is getting thicker.

Most importantly, I have noticed an internal change. I am now able to listen to **#21** all night long and have reached a level of internal peace and happiness that I have never before experienced.

In my work as a psychotherapist, I have experienced many different healing modalities. I believe that SWE is the fastest and easiest method for growth and healing. Clients have just begun to use the tapes. So far, three people report decreased sleep problems. Another client noticed that her skin problem cleared up, and night hot flashes disappeared.

Thymus

Ellen Eberhard, Ft. Collins, CO

I have been seeing my chiropractor for several years, and he has helped me over time to clear things at many levels. This last time I saw him, after listening to the **Thymus** tape for three weeks, he was quite surprised at the development of my thymus. He told me that the thymus is doing some work on the cellular level. Also, my body is changing at the DNA levels. Now that the thymus is being activated, the work we are doing can be deepened to the cellular level. I was surprised my doctor was able to determine that my thymus had developed so much, since my previous appointment. It is very exciting.

Allergies

Pat Lambert, New York, NY

I noticed a warmth and tingling in my legs and in the base of my spine when I listened to the **Chakra #1** tape. I tend to walk with my feet turned out, like a penguin. I felt spasms and burning in the muscles of my hip joint as these muscles seemed to pull my feet in a pigeon-toed manner, straightening them. By the end of the tape, this burning had turned to warmth and all the pain in my hips was gone.

I have asthma, and have been told by my acupressure/massage therapist that the "chi" from my kidneys and lower body rises up and floods my chest cavity. This robs my adrenal glands and intestines of "chi," hence my allergies. This also inhibits my absorption of nutrients and my adrenal/kidney functioning, leading to imbalance of my electrolytes and a depressed immune system. I could feel this energy being pulled down out of my chest, into my lower back and into the front of my body when I listen to the tape. How great to be able to feel the energy of my own body moving to the right place.

I should note that I've used the **Chakra #2** tape for relief of menstrual cramps. It works for three to four hours—then I play it again. It's certainly better than doses of pain killers or prostaglandins.

Lisa Mazurkewich, Stowe, VT

I ordered all four series of the tapes and had played different combinations for only about two weeks. It was time for our first Spring lawn mowing. We have a large, hilly lawn and this is quite a chore for me. In the past I would end up being sore in every muscle in my body for 2 or 3 days. The next morning I expected this to be the case. Amazingly, I woke up and realized that I did not have one sore muscle in my body. This was the first thing that happened to me with the SWE Tapes.

For the past 6-8 years I have had an ongoing problem with digestion. It got progressively worse, and I had a healer work on me. This helped a great deal and my pain was gone for some time. But due to my eating foods that I was not aware of being allergic to, I was still having some problems. My allergic reactions were severe stomach pains, severe headaches, problems swallowing, digesting and eliminating food. I had to go on a very strict diet and was unable to eat dairy products, wheat products, sugar, eggs, soy, tofu, bananas, peas, pineapples, breads, and grains (except corn and rice), chocolate, onions, garlic, fried foods, all preservatives and most spices, and any products containing any of these items. Needless to say, it was difficult trying to prepare a family meal. Three weeks after using the SWE Tapes I decided to eat out and test my response to the tapes. I ordered four items that I was not supposed to have and I did not have a single reaction to them. I was ecstatic!! I have continued to improve and I am now eating most foods. My digestion and elimination are normal now. What a tremendous relief all of this has been for me !

Other things have happened that I find very interesting. I saw the auras of the trees in my yard for the first time and this was exciting! I have been able to remember dreaming a lot more and my memory, focus and concentration is much better. My dowsing skills have improved and I am using the MerKaBa Tape in my MerKaBa breathing exercises. When I pick up the dowsing rods now it is as if they are motorized. The left one spins counter clockwise (masculine) and the right one spins clockwise (feminine) at a fast speed. So, I feel the activation of the MerKaBa has improved greatly. I love all of the tapes and I love saying the intentions. I would have to say now, that my favorite tape is the **Cell Rejuvenation** tape. I always feel good after playing this all night! Thank you for all of your work, Nicole

Arthritis

Barbi Lazonby, Gainesville, FL

This morning a Reiki client called me. She has rheumatoid arthritis and brittle bones, has been in a cast for two months as she cracked some bones and injured her knee, and was in severe pain. During her Reiki treatment visit, I played **Bones** along with **Chakra #3** tapes as she's low on self-esteem too. She was scheduled to have the fluid drained and a cortisone shot to the knee, yet when she went in to see her doctor, there was so much improvement, she did not need either!

Thank you for your wonderful work, and for allowing me to distribute these tapes. I'm very encouraged by the results both in my life and in my clients'.

Migraine

Michel Martel, Montreal, Canada

After a fall when I was infant, the headache that I had became continual, and by the age of nine, I had migraines – a minimum of one a week. These migraines would create nausea and make me very depressed. Sometimes it would be so bad, I would end up in hospital, where the doctors would inject a substance into my brain so that I could not feel anything. I was on anti-depressants, and more than one time, I tried to finish with this pain once and for all by trying to end my life. I did a lot of mental and spiritual work on myself to repair my brain cells and thought patterns. I always succeeded in controlling the pain, but still, I had the migraines and constant headache until I started to listen to the SWE tapes. After one month of working with all the SWE tapes, I stopped having those migraines, and the following month, not even a small headache anymore. Now, it has been six months that I AM FREE from pain and what is even more exciting is that I am motivated to accomplish the mission I had begun. I have the "JOIE de VIVRE," something that I had lost a long time ago. I feel great and I am forever grateful for the SWE tapes.

Susan Herzberger, Fairfield, IA

I want to tell you that I have experienced wonderful benefits from tape **#21**. After playing it for three days and increasing the time by ten minutes each day I became aware of what I needed to confront in my life to bring in more light. I was able to see the largest issue of my life which I never clearly saw before. Corresponding to this revelation, certain areas of my physiology were vibrated and my chiropractor said that bones had actually moved! Intense emotional catharsis occurred and I was easily guided to seek help which was tremendously effective in clearing and clarifying what needed to move on a cellular memory level. Now I am able to listen to that tape for over half an hour a day with much pleasure although I find that emotional issues of all sorts do not stay unresolved in my life for long. I consider Tape #21 a kind of truth serum for the courageous.

I have also had good fortune with the **Cellular Rejuvenation** tape. From the first time I played it, I could feel enliven in areas of the body which have been sources of pain in the past. My head, especially, really woke up and I have felt much happier to feel pleasure there, after having so many migraines in the past.

Pregnancy

I frequently hear from women that their menstrual problems have gone away, that their hot flashes relating to menopause have disappeared, and that some have even resumed their periods. It is certainly not pleasant to have menstrual pain every month. When periods are normal, without discomfort, however, we know that our glands are functioning normally and producing the hormones that give the body its youth, stamina and health.

Lynela Becker, Missoula, MT

One year ago, a young lady heard about the **SWE** tapes from her mother who had purchased them and experienced many benefits. The daughter had experienced three miscarriages so she was looking for something to help herself. Her doctor was not very enthusiastic about her becoming pregnant again.

She bought the tapes and started to play them every night. One month later, she became pregnant and nine months later, a beautiful and healthy girl named Christina was born. I spoke with the mother when Christina was nine weeks old, and the two of them were very happy and healthy.

Entities

Lara Niemoth, Canton, OH

I had temporarily moved to live with my great aunt for the summer. My bedroom was on the second floor of her house and we were the only two people living in the house.

I had great difficulty sleeping for the first two weeks I was there. I was feeling unsettled and disturbed when I was in my bedroom, but didn't know what to do about it until my mom suggested that I use the **SpaceLight** and **MerKaBa** SWE tapes. Nicole had said that "Nothing that is not of the Light can stay in the room once you have cleared it with the SpaceLight tape." So I played it.

That was the very first night since I had arrived that I slept peacefully and well. I got up in the morning after having a deep and refreshing sleep and walked down the hall to the second floor bathroom. I was startled to see that the bathroom had been disturbed. A terry cloth towel had been neatly folded across the top of the toilet tank, a small basket of artificial flowers and a can of Lysol had been sitting on top of the towel when I went to bed the night before. But now the flowers were dumped upside down and the Lysol can was laying on its side and most surprising of all, the terry cloth towel was on the floor all wadded up in a ball.

I went downstairs and noticed "icky" energy on the stairway as I descended to find my aunt. I asked her if she had come upstairs during the night or early morning for any reason. (I didn't want to freak her out.) She said, "No" and that she hadn't been on the second floor since I moved in.

Well that answer eliminated any people from the equation! I spent the next few days using the SpaceLight tape in the upstairs hallway and bathroom, the other bedroom, down the stairs and throughout the whole first floor. The entire house feels good now.

In addition, each night I sleep with the **MerKaBa** tape on autoreverse. I sleep like a baby. I have not had any "negative energy" problems since I cleaned the house with SpaceLight and protected myself with MerKaBa.

More Energy

Bonnie Lou Marceau, Wisconsin Rapids, WI

I have been using the SWE tapes for several weeks now. Each time I play a **Chakra** tape, I have noticed and felt energy moving in that chakra area. I also have more energy than usual and I sleep much better at night than I have in years! I am truly grateful for these tapes.

Thank you Nicole for your work and the sharing of it! Words are hard to find in expressing the gratitude I feel.

Judi Slaughter, Scottsdale, AZ

I am so grateful that the gift of your sound vibration tapes made their way into my life. Magic has been abounding in my life. My heart has been filled with joy and laughter. It seems that things I'm looking for, or questions I ask, are answered quickly.

Also, people are telling me I look younger than the last time they saw me. I look different. I know I feel different. I have noticed that I'm more aware of all the choices I have in my life. I want and find myself making choices that keep me in my joy. Thank you so much, Nicole, for sharing the gifts of the SWE vibration tapes. God Bless.

Autism

Minc, TX

I am using SWE tapes by Nicole La Voie and am noticing tremendous differences in… well, my husband and my son. So far I cannot attribute the change in Lacey's symptoms to the SWE tapes in particular, since I am doing so many new things simultaneously, but I have no doubt (since my boys are changing for the better as far as their attitudes, interest in spirituality, etc.). The frequencies are working toward balancing the chakras and promoting good health. In fact, Lacey told me so through the telepath the other day. She claims the SWE tapes make her body feel "less strange."

Minc is the mother of the most precious little girl who is affected by autism. One of the things Minc had mentioned, two weeks after I spoke with her on the phone, is that Lacey had doubled her vocabulary. However, it is difficult to say what did what, since Minc started using color as well as the sound frequencies. I encourage people who work with other modalities to continue to use what they feel works and add the sound frequencies. I believe that when we can approach a person from the different senses, we can only obtain better and faster results. The SWE tapes have demonstrated that they enhance any other technique or approach.

Classroom Behavior

Christine Richter-Brozowski, Newbury, OH

I needed to thank you in writing on behalf of myself, as well as my high school students. The tapes have wrought many wonders. On a personal note, I have been able to **free myself of a twenty-year reliance upon headache "remedies" and even the use of Premarin,** hormone therapy treatment that I was put on after a partial hysterectomy due to a fallen uterus from childbirth, years of running, and ensuing aging. I have never been one to run to medical doctors, having grown up under the regis of a homeopathic practitioner-father. So I had a lot of unease in beginning this hormonal treatment about six months ago. Then a friend recommended Nicole LaVoie's SWE tapes and her book to me. I read the book over a weekend, and on Monday, I personally talked with Nicole LaVoie over the phone. I knew when I heard her voice that truth spoke through her. I ordered the tapes. Two days later, I received my **Basic Series.** Three days later I cleaned out my cabinets of Excedrin, Premarin and Aspirin. I have seemed to get better and better: my eyes are clear, my chest is breathing freely again, my skin is clearer and lighter, my energy is almost boundless. And now I will share with you what I have done at my place of work.

I am a teacher for twenty-four years now. I had begun at the early age of twenty, teaching 7th through 12th grades, English, French and Theatre Arts. I have seen great successes and tragic failures within the student body. Over the years I have sought the best, most effective way to bring about positive results for my students. Results based on a way that would help them in their other classes, as well a kind of carry-over that would continue to benefit them.

I decided to bring the tapes to school and play them ever so softly during particular classes—especially for my last period English class of twenty boys and two girls. This was the class of "failures" – failures of previous eighth and ninth grade classes, as well as the state proficiency classes. Yes, the first nine weeks is in— they are passing all of their classes. Today they took the writing/ proficiency exam. We will know the results in December, but I can tell you that when I asked the kids what they did, how they did, and how they felt they had done—they were very confident.

From their sharing of their writing processes, I do believe they have, indeed, passed the state-mandated exams. The tapes of **Clarity and Focus, Brain, Courage and Prosperity**, as well as **Merkaba** were played almost continuously during the 45-minute class times. My students even ask me to play the "energizers" which is what I call them. Nicole, thank you for these miraculous tapes. They have changed our lives.

P.S. By the way, my lips are even fuller! A teacher friend asked me if I had had a collagen lip implant. Wow!

Drug Addiction

Mario Perazzelli, Montreal, Quebec, Canada

Nicole, the tapes you gave me for the work I did for you did wonders for me. At first I did not believe in them and I didn't really listen to them (I tossed them aside in my wall unit!!!), but NOW, I think that they saved my life.

You know that I had problems with a lot of things in my life; I was a manic-depressive, I had to take Lithium to control my mood swings. I even lost two jobs because of that condition. I didn't like the effect of the Lithium pill so I decided to go on street drugs (Cocaine, Marijuana, and others). I was also a self-centered, egocentric being – I thought I was the only person deserving to live on this planet, I had a lot of anti-socialism reaction and I was becoming very aggressive and negative. My own mother was afraid of me.

Then you called me, on August 1997—a Thursday night, I still remember – to ask some news about my condition. Following that long discussion, that's when I really started listening to the tapes. After a period of three weeks only, I was out of the drugs, without any withdrawal symptoms. I was in control of my life. That is when I called you back to ask if you needed people to help you in your work. Then I came to work with you (and I am still), and I have used the tapes since then. After going through all those tapes, I can tell you that my life has changed in unbelievable ways. Now, I respect others, I am in love with nature, I am no more judgmental of people, and—something I would have never imagine—I don't bite my nails anymore (that is almost incredible for me!!!).

Anyway, I would just like to tell you that in all my years of knowing you, I would never have thought that you could help me the way you did. You really are on a mission to help others. I am very happy to be a part of this new technology, and at the same time, I am very eager to share my personal experience with others.

Sensory Integration Dysfunction

Jean Worth, Washington DC

Paul, my 5 ½ -year-old son, has Sensory Integration Dysfunction. He can't regulate what comes in, and when he becomes overloaded, he can't regulate his behavior or output. This can come on with no advanced warning (signs), and leave in its wake a shattered little boy. We live "on alert," and it has been a very rough five years.

His teacher from the Washington Waldorf School, Carolyn Robinson, says: "When Paul entered my class in the fall of '97, he resembled a tornado. Paul would run around the room leaving a disastrous trail behind him. He would knock over a block structure, take someone's marbles, scatter a well planned dinner in the doll house, and drive into someone's cozy house that they had just built with clothespins and sheets. No one wanted Paul near their play area.

"Paul was evaluated by our school's remedial teacher. A recommendation for Paul was to do Eurythmy and Sensory Integration Therapy. Paul continued not to come when called, and would run from the teacher. Other children would still not play with him.

"Sound Wave Energy was introduced and a big change was noticed in Paul. Paul was responding when we called his name. He looked us in the eyes for the first time. He was able, on his own, to enter into play with the other children. Paul is able to come into the classroom and start his day without running around the room three times. He is able to sit and do finger knitting and is very proud of what he has made. Paul is now able to be a friend."

Cheri Present is an O.T. in Washington, D.C., at Lynn Israel & Associates. "The first time I saw Paul, he was extremely disorganized. He couldn't filter out any information, couldn't regulate his voice, couldn't be touched, he made no eye contact and displayed random, unplanned movements throughout the entire session. When Paul walked in two weeks later for the third session, he looked like a different person. He was calm, his voice was regulated, he made eye contact and carried on dialogue throughout the session. When I spoke with his mother, he calmly waited his turn. His motor activity was well planned and steady. I was left feeling amazed. This was not the same child I had encountered two weeks earlier. Of course, I was curious about what was causing this dramatic change, and was told by his mother that they had started using the SWE frequencies just four days prior to the meeting. I am eager to know more about this work!"

Well, these are comments from two very important people, who want to help Paul build skills for having a happy and successful life. What I noticed at home was also profound. Within 48 hours of playing the tapes, Paul slept through the night for the first time in his life. He got hungry and tired for the first time in his life. He gained three pounds in three weeks and started resting and napping. He can actually sit down and look through a book and can play alone for extended periods of time.

Paul's energy still abounds with curiosity, humor and the love to match. The difference is, that now he is more regulated and we can enjoy more frequently all of his good qualities.

I believe so deeply that the SWE Tapes have given us, and our little boy, a new lease on a new life together. I remain eternally grateful to God for giving Nicole her gifts. I shall be forever grateful to Nicole for sharing her gifts with us.

Sleeping and Craving
Cathy Littlefield-LeVasser, Guided Imagery Facilitator
Certified Sound Wave Energy Practitioner, Fairfield, CT
I have been using the frequencies for five weeks and was surprised at all of the subtle changes that had occurred when I checked my

evaluation sheet. My cravings for chocolate have disappeared. I suffer from Fribromyalgia, sleeping is a real problem for me. I am sleeping deeper than I ever slept since the disease developed. In fact I sleep so deeply that my husband has been checking to see if I am breathing at night! I can hardly wait to see what happens with continued use of the frequencies.

SoulNote

Joanie Martin, Akron, Ohio

After waiting for two years to get my **SoulNote** I was both apprehensive and excited about the opportunity. Actually, I had no expectations for this experience, which was good because in my most creative dreams I could never have created this incredible experience.

Even after using the tapes successfully for two years, I approached this new experience with skepticism. This vanished very quickly. From my reading and talking with a friend I knew the beginning process was to talk about something I felt passionate about, something that came from my heart. Prior to our session, several topics were considered and outlined in my mind. What actually happened was in the moment. Spirit moved me and I really spoke from my heart and any preparation was wasted time.

One of the most amazing things about this experience was how a change in frequency would elicit changes in my feelings. Through this process my emotions changed from completeness, to sadness and fear, to uncontrollable joy, to the safety and love of God.

Vivid mental images often accompanied the feelings. These pictures come back to me as I play the tape and even during my day they return and bring me great comfort.

When Nicole determined my SoulNote and played it for me *I felt like a part of me had been found.* There was a sense of completeness. Now, when I hum it without the tape, I know exactly when I have the right tone. My whole body resonates with it, and then comes the sense of peace and completeness.

The lower frequency truly connected me to some past earthly experiences. Feelings of fear and sadness were very strong. When

Nicole used the releasing technique I felt 50 pounds lighter and freer.

The double frequency of my SoulNote brought me incredible joy. Giggles were difficult to stiffle. Laughter came from deep within. True joy just seemed to bubble up. My spirit guide has been telling me for several years to focus on the joy. The highest tone Nicole gave me was a very special gift. *With this tone I could feel God's love.* I could actually feel his arms around me. It was spiritual on a very deep level. For the first time on this earth plane I got it. I got a sense of the oneness. This feeling continues to stay with me. Now in my meditations I often feel the physical presence and feel the love. When the session was over, my dog and I went hiking. Everything glowed—the trees, the plants, the rocks. Everything. There was a lightness and joyfulness in my body that wasn't there before. I was happy on a level deep within me.

Every day I play my SoulNote as I'm driving to and from work. It has become an instrument for me to assess how life situations have affected me. When I experience situations and people as good, my SoulNote feels good. I hum it with ease and can feel the resonance easily. When I haven't been true to my spirit and when the world has affected me negatively finding the right tone is more difficult and takes more effort. At these times the tone is not comforting, it is an irritant to my ego self. I persist in playing it and slowly find myself returning to the state of peace and lightness you've showed me how to find.

You truly have a wonderful gift for the world. Thank you for sharing it with me.

Testimonials from Merily Preston, VA

Janice Miano

Janice was considering surgery for carpal tunnel syndrome. While using the Sound Wave Energy tapes, however, Janice has no pain, even when she works on her computer for long hours or does remodeling in her store, The Angel Shop in Occoquan, VA. When the tapes are not being played, Janice feels the pain again.

Ruby Geisel

Ruby had been feeling pain from a chronically swollen ankle. Pain medication for a day and a half had brought no relief. During our lunch at a restaurant, Ruby had not mentioned the pain to me. Afterward, as we had tea at my house, I played the **Circulation** tape. After 20 minutes, she raised her hand to stop me from talking. Ruby said that her ankle no longer hurt. She couldn't believe it. Her ankle never returned to the bad shape it was in on that day, although it is still swollen. She now has her own SWE tapes.

Marsha Hill

Marsha had a severely painful case of gout in her big toe. I started her on SWE tapes and saw her two weeks later walking normally and wearing shoes. I asked her what happened to the gout. She said that the SWE tapes had taken care of it. Once the SWE therapy began, she abandoned her conventional treatments.

Melissa Preston

Melissa, my daughter and a graduate student, was diagnosed with probable MS. She made many positive changes in her life to deal with this possibility, including meditation, regular exercise, a healthy vegetarian diet, and a change in career plans. She also played my SWE tapes. After several weeks listening to the tapes, she announced that she no longer has MS! Her latest tests confirm that she is now at low risk for having the disease. She decided to purchase her own set of tapes after waking one morning wanting to play the **Purification** tape. She feels that playing her **Basic Series** has enabled her to make decisions from a higher level and to clear many emotional issues.

Heaven on Earth *13*

Return To Harmony

Sound Wave Energy was founded in 1993. This work is the culmination of eight years of study and exploration in related fields. The specific information that made the creation and production of the special, low-frequency SWE audiotapes possible was received from Spiritual Guidance, including instructions for the ideal usage of the tapes. This Guidance also included a development plan for establishing a Center of Sound and Light.

SWE has grown steadily since its 1993 inception, through personal contact and word-of-mouth referrals by over 3,000 clients who have experienced wonderful results using the tapes. SWE recordings are helping people in all fifty-two states, Canada, Central and South America, seven European countries, three Asian countries, Australia and New Zealand. In fact, they are in use on all continents, with the exception of Africa, at the time we go to press.

Based upon this enthusiastic response, a business plan has been developed to guide expansion of the existing business. This expansion is intended to help finance The Center of Sound and Light in Southwest Colorado or elsewhere. An Awareness Center is envisioned as a focal part of a multi-purpose center which will be staffed by those in residence in the center community. The low frequency sounds will play a central role in setting the atmosphere of the Awareness Center, as well as the entire center, to prepare people for Ascension.

The word ascension, as used in the SWE context, is not intended in an elitist or escapist connotation. In SWE vernacular, ascension connotes grounded people who joyfully serve Earth and her residents while raising their vibrations into the octave of Oneness. Their behavior and attitudes will be consistent with the principle of unconditional love as they go about living their lives here on Earth. Joy will be the normal state.

It is Sound Wave Energy's mission to facilitate the ascension of humanity, by promoting the growth of each individual's commitment to God and humanity in accordance with the plan of Earth's Spiritual Hierarchy.

The Goals of Sound Wave Energy

One of our primary goals is the greater distribution and use of SWE recordings world-wide. These special CDs and tapes provide the means for individuals to take responsibility for their personal growth and to prepare themselves for higher vibrations and ascension on Earth.

In order to achieve this goal, it is essential that those wishing to be distributors of SWE frequencies have extensive experience with using the recordings for their own benefit. This is an absolute prerequisite. Distributors will share the common thread of setting intentions and using the SWE recordings in order to accelerate personal growth and preparation for ascension. They will assist others to prepare for the shift into the fifth dimension, expected in the next twenty years.

SWE therefore intends to inspire all SWE distributors to become balanced and to develop their full potential. This could include preparing to establish facilities in a variety of geographic locations which would be similar to the Center of Sound and Light. It is our intention that this effort be world-wide.

Global Harmony Marketing Plan

Sound Wave Energy has created a unique method of distributing the recordings. The Global Harmony Plan has been designed to allow those who want to share the benefits they have received with the SWE recordings to also share the profits SWE will receive from the sales of the recordings. To be a part of this Global Harmony plan, an individual will need training in the holistic approach, the theory of wellness, the spiritual, mental, emotional and physical aspects of our beings, and how the recordings work with these aspects. The training is based on co-operation and unity of purpose. Trainees are given an opportunity to start a personally fulfilling career that allows them to make a difference in the world, sharing Light, Love, Health and Wealth. To do our part in implementing these energies on the planet, we have organized a global participation to raise humanity's awareness.

Global Listening

At the same time each day, at various locations, we have been listening to the same SWE recording. The goal of this listening is two-fold: 1) SWE recordings raise our vibrations at the spiritual, mental, emotional and physical levels, and 2) when the same vibrations are played at the same time in many locations, with a special intention, it is as if thousands of lights are transmuting the darker energies into Love and Light. These energies are then available for all who are ready for transformation. Together we will saturate the grid with loving energy so that everyone on the planet can benefit from it and access it when needed. The more people use the frequencies, the easier and faster the level of consciousness of the planet will be raised!!!

We can say The Great Invocation (see page 266) at the beginning of the listening period. If we want to invite other people to listen with us, it is a great idea. We can also listen by ourselves, knowing that thousands of others are doing the same, focusing on sending these vibrations into the grid surrounding the planet. The objective is to do it together to create a vortex of Light and Love around the planet.

Imagine many people, all over the world, playing the same frequencies, with the same intention, all at the same time. Because each person has free will, we don't want to "send" these frequencies to specific people if they haven't asked for them. However, if our intention is to send them into the grid system surrounding the planet, and have them available when a soul reaches for them, then we do not interfere, and we still make them available to everyone.

The Great Invocation

From the point of Light within the Mind of God
Let light stream forth into the minds of men.
Let Light descend on Earth.

From the point of Love within the Heart of God
Let love stream forth into the hearts of men.
May Christ return to Earth.

From the centre where the Will of God is known
Let purpose guide the little wills of men –
The purpose which the Masters know and serve.

From the centre which we call the race of men
Let the Plan of Love and Light work out
And may it seal the door where evil dwells.

Let Light and Love and Power restore the Plan on Earth.

Training Distributors of the SoulNote

The SoulNote is the physical representation of our state of being. That's why when we hum our note, we feel in total harmony. When we find our SoulNote and use it, we connect with our inner being, and allow the soul to move toward more and more communion with our inner self. Finding one's SoulNote is so important, and yet we have been limited because it must be done in person, one-on-one. Over the years I have been the only one testing for SoulNotes. Now a program is underway to train individuals to use the computer to identify SoulNotes, so that more and more people may find their note, and return to harmony.

Funding

We are in the process of raising funds to acquire the perfect building site for the Center of Sound and Light. Once we are guided to the ideal location, further funding will be needed in order to construct the necessary buildings. An Awareness Center, wellness center, research and conference center, art center, gymnasium, community housing, guest lodge and service buildings are among those planned.

SWE is presently seeking financial support from conventional and alternative philanthropic sources. In addition to a donation of any amount, contributions of land, materials, professional expertise, and personal assistance will, of course, be gratefully welcomed.

It is important to know that the center has a *non-profit and tax-exempt* status. All contributions, grants and donations to the Center of Sound and Light will receive a receipt in order to document a tax deduction. If your heart resonates with this, you are invited to send your donation to the address listed in Product Information at the back of this book.

We intend to provide benefactors with more than tax writeoffs. Gifts of SWE recordings and complimentary time at the center will be included. Benefactors will have a wide variety of choices available to them to create a transformative experience in this beautiful place, a true heaven on earth.

Once construction reaches a functional state of completion and we begin providing services, it is anticipated that the daily operation of the center will pay its own way. Of course, all subsequent revenue from the sale of books and recordings will continue to support the operation of the center.

Land

Acquisition of the ideal land for the Center of Sound and Light is the next step. We wish to attract the cooperation of an experienced developer to work closely with us to plan the complex. In addition, we intend to attract other key people who will resonate with the spiritual vision underlying the entire project.

Community

The plan includes the establishment of a *self-sufficient Community* of like-minded people who wish to live in harmony with the Spiritual Hierarchy, each other, nature, and to assist in the work of the facility. These community members will be drawn together by the idea of becoming living examples of those preparing for ascension by learning the basic laws of the Universe that were once taught in the Mystery Schools. This community of like-minded people will live in self-sufficiency and harmony with Mother Earth.

Heaven on Earth for me is real right now and I know it will be even more so in the near future. For me, ascension means bringing down the I AM into my body here on Earth and not going anywhere until my divinely ordained transition back into Spirit. I know others will resonate with this and that together we will co-create the community and the Center of Sound and Light. Together we will enjoy working to help the guests and ourselves in loving and joyful ways.

Additional Building Plans

Plans also include special housing for staff, guests and visitors plus campsites and additional facilities for maintenance, food production, service and other support activities.

We intend to build a Research and Conference Center whose purpose will be to measure and document the results of the ongoing application of developmental tools and to establish the credibility of the entire program. Findings will be used to establish a bridge between the spiritual and scientific communities. New discoveries will be shared via a flow of people coming to speak to us of their new found gifts.

Naturally we plan to build a Wellness Center to enhance physical, emotional and mental growth. The best techniques available will be offered in this center, provided by skilled and loving people. Activities at the Wellness Center will prepare people to enter the Awareness Center, the ultimate place for total transformation.

The Awareness Center

The Awareness Center will be an essential aspect to the whole center. It will be used to prepare individuals for their transition into a fully spiritual society. This Awareness Center will not be a typical place of worship (such as a church, temple or mosque). There will be no religious affiliation, dogma, or practices. No intermediary will be needed nor involved. Instead, the Awareness Center will be a place to individually and collectively re-establish direct personal contact with our Higher Self, our Spiritual Hierarchy and with the Creator. The work done in the Awareness Center will accelerate the recognition of and honoring of the God Within. The use of sound frequencies, light, color, crystals, geometric shapes, water and aroma will be among those tools used to raise ourselves to higher levels of spiritual awareness.

In my vision, the Awareness Center will be designed of six round chambers arranged in a circle and forming a star tetrahedron. Other sacred geometrical forms will be used in the structure of these chambers and in the crystal-domed atrium covering the central gardens. Beautiful marble and crystal support pillars will emanate rainbow colors. The corresponding sound, light, color, crystals, aroma and water will be separately arranged in the different chambers.

It is hoped that there will be many people coming together to enjoy this very special place. Our intention, in creating this sacred space, is to facilitate the ascension of humanity by promoting the growth of the individual's commitment to God and to human service in accordance with the plan of Earth's Spiritual Hierarchy.

Appendix

Intentions

INTENTIONS FOR THE FOUNDATION SERIES

Chakra #1 - (Energy Center #1)

I AM grateful for the easy and graceful transformation that is occurring in all of my bodies as a result of using these frequencies.

I AM absorbing the frequencies of the first Energy Center, or Chakra #1, into my system. *I AM* visualizing the shape of the cube and the color red. *I AM* secure and safe. *I AM* in my body. *I AM* balanced in my first Energy Center. *I AM* here to spiritualize my body. *I AM* letting go of whatever no longer serves the *I AM* that *I AM*. *I AM* abundant: *I AM* healthy. *I AM* clean and rejuvenated because my elimination system works well. *I AM* creating a vacuum that attracts everything I do need. *I AM* assimilating, through these frequencies, all the nutrients my body needs. *I AM* preparing to awaken my Kundalini, so when it is the right time, this energy can rise harmoniously through all my Energy Centers and bring me enlightenment! *I AM* achieving perfect balance in my first Energy Center easily and gracefully. SO BE IT! Only that which is for my higher good can come from using these frequencies.

Chakra #2 - (Energy Center #2)

I AM grateful for the easy and graceful transformation that is occurring in all of my bodies as a result of using these frequencies.

I AM absorbing the frequencies of the second Energy Center or Chakra #2 into my system. *I AM* visualizing the shape of the pyramid and the color of orange or pink or whatever color seems appropriate to for my second Energy Center. *I AM* energizing my spleen, my reproductive organs and the oxygenation of my body. *I AM* aligning my masculine and feminine aspects to create the energy of perfect gender balance. *I AM* in a loving relationship with myself and others. *I AM* perfectly balanced in my kidneys, pancreas and other organs related to this area of my body. As *I AM* resonating with these frequencies, a new sense of appreciation and love for my masculine and feminine aspects comes into my conscious awareness, so that balance and integration of these two vital aspects is achieved. *I AM* now assimilating all the nutrients which are needed to achieve perfect balance in the second Energy Center. *I AM* achieving perfect balance in my second Energy Center easily and gracefully. SO BE IT! Only that which is for my higher good can come from using these frequencies.

Chakra # 3 - (Energy Center #3)

I AM grateful for the easy and graceful transformation that is occurring in all of my bodies as a result of using these frequencies.

I AM absorbing the frequencies of the third Energy Center or Chakra # 3 into my system. *I AM* visualizing the shape of the sphere and the color of yellow or green or whatever color seems appropriate for my third Energy Center. *I AM* in balance in my third chakra. *I AM* in power in my life. *I AM* responsible. *I AM* a powerful being of love, harmony, balance, joy, happiness, and abundance. *I AM* perfectly digesting, absorbing and assimilating food, gifts and life's challenges. *I AM* easily assimilating changes so that I may be able to receive the gifts of Spirit. *I AM* honoring my feelings so that I may achieve balance and resonance from my inner guidance as my third chakra is balanced. *I AM* directing the energies of my solar plexus center to my heart center where my Power of Love abides so that the desires of my little self will be transmuted into service to my Higher Self. *I AM* achieving perfect balance in my third Energy Center easily and gracefully. SO BE IT! Only that which is for my higher good can come from using these frequencies.

Chakra # 4 - (Energy Center #4)

I AM grateful for the easy and graceful transformation that is occurring in all of my bodies as a result of using these frequencies.

I AM absorbing the frequencies of the fourth Energy Center or Chakra #4 into my system. *I AM* visualizing the shape of the cross and the color green or gold or whatever color seems appropriate to my fourth Energy Center. *I AM* easily assimilating the frequency which will accelerate the opening of my thymus. *I AM* absorbing these frequencies so that my heart and thymus now become one, allowing my human conditional love and UNIVERSAL LOVE to unite into ONE EXPRESSION of Pure Reason. *I AM* responsible for myself, my beliefs and my feelings. *I AM* cleansing my body through the perfect purifying function of my heart, lungs, circulatory system and lymphatic system. Through the use of oxygen, *I AM* efficiently eliminating unhealthy particles. *I AM*, through perfect assimilation of these frequencies, rejuvenating my body and strengthening my immune system. Illness, aging and death no longer touch me. *I AM* using the frequency of the growth hormone in Chakra #4 for ongoing healthy etheric growth as well as for infinite spiritual growth in service to humanity. *I AM* achieving perfect balance in my fourth Energy Center easily and gracefully. SO BE IT! Only that which is for my higher good can come from using these frequencies.

Chakra # 5 - (Energy Center #5)

I AM grateful for the easy and graceful transformation that is occurring in all of my bodies as a result of using these frequencies.

I AM absorbing the frequencies of the fifth Energy Center or Chakra #5 into my system. *I AM* metabolizing these frequencies perfectly in cooperation with the rest of the chakra / endocrine system. *I AM* visualizing the shape of the crescent and the color blue or whatever color seems appropriate for my fifth Energy Center. *I AM* absorbing these frequencies that balance my thyroid and parathyroid glands which normalize and stabilize my metabolic rate. *I AM* acknowledging both my masculine and feminine aspects. *I AM* living my life guided by my heart and love. *I AM* in harmony with myself and allowing my feeling of openness to increase the clarity of my speech. *I AM* resonating with the truth within me. *I AM* speaking my truth all the time. *I AM* achieving perfect balance in my fifth Energy Center easily and gracefully. SO BE IT! Only that which is for my higher good can come from using these frequencies.

Chakra # 6 - (Energy Center #6)

I AM grateful for the easy and graceful transformation that is occurring in all of my bodies as a result of using these frequencies.

I AM absorbing the frequencies of the sixth Energy Center or Chakra # 6 into my system. *I AM* visualizing the shape of the tetrahedron and the color Indigo or whatever color seems appropriate to this Energy Center. *I AM* open and activated in my third eye and pituitary gland as I absorb these 'knowing' frequencies. *I AM* opening a window to my past incarnations, allowing higher knowledge and experiences that are relevant to my current life situations to come into my awareness. *I AM* vitalized in my eyes, ears, nose, teeth and lower brain, allowing these aspects of my physiology to be cleared and raised to their highest frequency. *I AM* balanced in my production of acid / alkaline, which leans perfectly towards alkalinity. *I AM* harmonized and balanced in my brow chakra, and thus my telepathic powers become more developed. Because I want to see all my vision is perfected. *I AM* achieving perfect balance in my sixth Energy Center easily and gracefully. SO BE IT! Only that which is for my higher good can come from using these frequencies.

Chakra # 7 - (Energy Center #7)

I AM grateful for the easy and graceful transformation that is occurring in all of my bodies as a result of using these frequencies.

I AM absorbing the frequencies of the seventh Energy Center or Chakra # 7 into my system. *I AM* visualizing the shape of the Lotus blossom and the color purple. As I absorb the frequencies of Chakra # 7, *I AM* including the balancing and availability of Chakras # 8 & 9. *I AM* clearing out any blockages so that the energy may flow freely, smoothly and harmoniously throughout the physical and etherical bodies. *I AM* aligning and balancing this chakra with my entire chakra system and all associated endocrine glands, activating the pineal gland. *I AM* achieving perfect balance in my seventh Chakra easily and gracefully. *I AM* in perfect balance and harmony, and energy flows in each of my chakras. *I AM* accelerating my connection with DIVINITY. I KNOW WHO *I AM*. SO BE IT! Only that which is for my higher good can come from using these frequencies.

Love

I AM grateful for the easy and graceful transformation which is occurring in all my bodies as a result of using these frequencies.

I AM feeling an abundance of Love. *I AM* listening to these frequencies with a feeling of anticipation, joy, and wonder. *I AM* benefiting from these frequencies in my heart muscle, lungs and arteries as the physical pumping muscle responds by becoming stronger, and more open, as do my feelings. *I AM* creating balance between my heart and my brain so that my decision making can be the result of perfect collaboration and cooperation of each. *I AM* experiencing unconditional love. . .the impersonal love that loves everything with grace and non-judgement. *I AM* giving and receiving unconditional love. *I AM* open, expansive and joyful. I know that Everything is Perfect. . . Everything is in Divine Order. SO BE IT! Only that which is for my higher good can come from using these frequencies.

Relaxation & Calmness

I AM grateful for the easy and graceful transformation that is occurring in all of my bodies as a result of using these frequencies.

I AM giving and receiving love. *I AM* relaxed and receptive to unconditional love. *I AM* letting go of my resistance and creating the opportune atmosphere for peace and change. *I AM* welcoming change with JOY. *I AM*, with the help of these frequencies, reminding my DNA cells of their original intent. *I AM* drawing my DNA cells back into alignment so that they will again provide access to all knowledge contained within its structure. *I AM* enjoying a sense of peace and well-being as I reassess and redefine my goals. *I AM* now able to become aware of and take charge of my heart's desires. I know that Everything is Perfect. . . Everything is in Divine Order. SO BE IT! Only that which is for my higher good can come from using these frequencies.

Brain, Courage & Prosperity

I AM grateful for the easy and graceful transformation that is occurring in all of my bodies as a result of using these frequencies.

I AM aware of minor adjustments in each wave form in my brain that take place so that my ability to make decisions is perfected in terms of the highest good for all concerned. *I AM* facilitated in my process of changing old beliefs that need to go. *I AM* in control of my thoughts. *I AM* grounded. *I AM* integrating the left and right sides of my brain. *I AM* able to have the endurance and courage to focus on my Self. *I AM* recognizing my birthright of abundance as the natural law of the Universe. I know that Everything is Perfect. . . Everything is in Divine Order. SO BE IT! Only that which is for my higher good can come from using these frequencies.

Higher Consciousness

I AM grateful for the easy and graceful transformation that is occurring in all of my bodies as a result of using these frequencies.

I AM releasing fears and doubts rooted in old belief systems *I*

AM releasing any disharmonic focus I may have on this plane. *I AM* gaining the benefit of feeling connected to each and all. I know that *I AM* a part of the UNIVERSAL SELF. *I AM* realizing my ONENESS. *I AM* accepting with LOVE whatever understanding assistance I may need to arrive at this place of Light. *I AM* raised to Higher Consciousness. I know that Everything is Perfect. . . Everything is in Divine Order. SO BE IT! Only that which is for my higher good can come from using these frequencies.

INTENTIONS FOR THE SPIRITUAL SERIES

Thymus

I AM grateful for the easy and graceful transformation that is occurring in all of my bodies as a result of using these frequencies.

I AM activating my Thymus gland. *I AM* immune to dis-ease. *I AM* immune to all negativity. *I AM* experiencing the rejuvenation of my body. *I AM* experiencing UNCONDITIONAL LOVE for myself and others. *I AM* absorbing the frequency of DHEA, the precursor of hormones, now the most abundant hormone in my blood stream. *I AM* claiming my immortality. I know that Everything is perfect. . . Everything is in Divine Order. SO BE IT! Only that which is for my higher good can come from using these frequencies.

Transition

I AM grateful for the easy and graceful transformation that is occurring in all of my bodies as a result of using these frequencies.

I AM accepting new things, new changes that are coming. *I AM* accepting miracles in my life. *I AM* asking for and receiving Miracles in my life. I honor MYSELF. *I AM* utilizing my will to transform energy to a higher frequency which moves me into change easily and gracefully. *I AM* now unifying at the level of the Self with the Almighty. *I AM* dissolving all resistance to love and to change. *I AM* one who uses physical experiences to the fullest. *I AM* experiencing the creative flow of energy. *I AM* resonating with these frequencies because I already have the connection with my inner guidance. I know that Everything is Perfect. . . Everything is in Divine Order. SO BE IT! Only that which is for my higher good can come from using these frequencies.

MerKaBa

I AM grateful for the easy and graceful transformation that is occurring in all of my bodies as a result of using these frequencies.

I allow my emotional and mental bodies to create around me, this vehicle of light that keeps me centered in my totality. *I AM* safe and secure in my Sacred Space so that I may readily contact my Higher Self. *I AM* experiencing peace and serenity as the MerKaBa frequencies create a spinning effect around my body. *I AM* residing in a space of Love, Light and Joy. I know that Everything is Perfect. . . Everything is in Divine Order. SO BE IT! Only that which is for my higher good can come from using these frequencies.

Energy, DNA & Enzymes

I AM grateful for the easy and graceful transformation that is occurring in all of my bodies as a result of using these frequencies.

I know my own BEING. I know the blood flow process. *I AM* conscious of the interaction between my brain and my higher self. *I AM* manufacturing and utilizing enzymes perfectly. *I AM* experiencing new strength in my pancreatic cortex. *I AM* having a deeper understanding of the shift of consciousness that can be manifested internally by brain wave states. *I AM* experiencing a strengthening of my digestive system as chakras 2, 4, 6, and 8 are aligned and in focus. *I AM* energized in my physical body, my digestive system and spleen. *I AM* able to metabolize lactic acid and other by-products of my muscular system. *I AM* releasing parasites, viro- and micro-organisms. *I AM* receiving help through these frequencies, with the transference of DNA information. *I AM* attuned to those of preceding generations. *I AM* having an easier time of attunement, conscious acceptance, and willing connection. *I AM* free from all auto-immune difficulties. *I AM* aware of all my bodies awakening in different areas. I know that Everything is Perfect. . . Everything is in Divine Order. SO BE IT! Only that which is for my higher good can come from using these frequencies.

Patience

I AM grateful for the easy and graceful transformation that is occurring in all of my bodies as a result of using these frequencies.

I AM assimilating new concepts and seeing things in a new manner. As a traveler, *I AM* adjusting easily to new time zones and new cultures. *I AM* enjoying a new relationship with Mother Earth. *I AM* empowered and strengthened by this relationship as *I AM* able to understand Mother Earth as a loving Being. *I AM* committed to using her energy more appropriately and lovingly. *I AM* feeling Earth's energy strengthening all my bodily processes, my digestion and my elimination. *I AM* gaining an awareness of Earth's geopathic energy for use in diverse situations, particularly dowsing. *I AM* able to receive an enlightening consciousness or awareness from the plant kingdom. *I AM* able to have a deeper sense of patience as I view things from the perspective of the plant kingdom.

I AM free from repeating secret dramas. *I AM* free from the emotional attachment to and the need for approval and attention derived from these dramas. *I AM* now perfectly grounded on Mother Earth. *I AM* experiencing stimulation in my nervous system so that it functions at a higher rate, enhancing my spiritual capacities and strengthening my ability to process and clear energies such as fear. *I AM* receiving benefits for my lungs from these frequencies. *I AM* now recognizing, processing and clearing all energies. I know that Everything is Perfect. . . Everything is in Divine Order. SO BE IT! Only that which is for my higher good can come from using these frequencies.

Oneness

I AM grateful for the easy and graceful transformation which is occurring in all my bodies as a result of using these frequencies.

I AM experiencing feelings of infinity and Oneness. *I AM* becoming open to the LIGHT. *I AM* communicating with all kingdoms on Earth and throughout the Universe. *I AM* united with all other dimensions of the Light. *I AM* releasing all judgments. *I AM* bringing the Oneness within myself. *I AM* merging with ALL THAT IS. I know that Everything is Perfect. . . Everything is in Divine Order. SO

BE IT! Only that which is for my higher good can come from using these frequencies.

Purification

I AM grateful for the easy and graceful transformation which is occurring in all my bodies as a result of using these tapes.

I AM able to release unconsciously gathered material entities. Now, without these entities, *I AM* able to grow and to take in energy from the environment, from food, air and water. *I AM* able to assimilate the helpful energies of the sun. *I AM* experiencing an energizing, a cleansing and clearing, and oxygenation of my blood. *I AM* recovering from my stay in hospitals. In all situations, *I AM* healing perfectly. *I AM* delighting in my perfect acid/alkaline balance. *I AM* able to pour energy into any situation. *I AM* able to increase my focus, to increase my group activities and to help levels of understanding in the world. *I AM* in the understanding that what manifests in my world comes from my Being first, then Doing, then Having. I know that Everything is Perfect. . . Everything is in Divine Order. SO BE IT! Only that which is for my higher good can come from using these frequencies.

Gateway

I AM grateful for the easy and graceful transformation that is occurring in all of my bodies as a result of using these frequencies.

I AM experiencing a constant rise in my vibratory rate. *I AM* opening my heart to allow the energies to flow into my body, coming from my *I AM* Presence and above. *I AM* integrating these new energies smoothly and gracefully. *I AM* anticipating with joy, my detachment from all that no longer serves me. I know that Everything is Perfect. . . Everything is in Divine Order. SO BE IT! Only that which is for my higher good can come from using these frequencies.

Immortality

I AM grateful for the easy and graceful transformation that is occurring in all of my bodies as a result of these frequencies.

I AM enjoying a high energy level as a result of increased stimulation of the endocrine function in my pineal and pituitary glands. *I AM* fully energized because my thyroid, thymus and adrenals function optimally. *I AM* in a state of healing and resonance in my pineal and pituitary glands. *I AM* breaking out of old patterns. *I AM* thinking in new ways. *I AM* experiencing cellular regeneration in a youthful manner, as a result of greater absorption of Xenon from the earth's atmosphere. *I AM* growing into new patterns of emotional understanding. *I AM* understanding the many and the one through inner coordination. *I AM* experiencing inter-dependency within my own body. *I AM* receiving a higher loving understanding and connection. I know that Everything is Perfect. . . Everything is in Divine Order. SO BE IT ! Only that which is for my higher good can come from using these frequencies.

Connections

I AM grateful for the easy and graceful transformation that is occurring in all my bodies as a result of these frequencies.

I AM open to my Soul Energy. *I AM* increasing my healing ability. *I AM* delighting in my healing light as it moves down through my arms and into my hands. *I AM* recognizing my inner light. *I AM* finding new pathways, new ways in which to bring my healing light to the World. *I AM* feeling a connection to another level. . .a parallel Universe outside the realm of our time and space barriers. *I AM* infused with forgiveness of self, clarity, awareness and understanding. *I AM* receptive to spiritually developmental information and extraterrestrial sources. *I AM* experiencing a strengthening of my psychic abilities and my higher God connection. *I AM* connected with the angelic realm and can easily communicate with them. *I AM* sharing and working with spiritual information in a proper way. *I AM* attuned to a sense of ease and an awareness of flow. *I AM* connected with my Soul. I know that Everything is Perfect. . . Everything is in Divine Order. SO BE IT ! Only that which is for my higher good can come from using these frequencies.

Shifting Consciousness

I AM grateful for the easy and graceful transformation which is occurring in all my bodies as a result of using these frequencies.

I AM becoming attune to greater humanity's purpose. *I AM* accessing a larger collective consciousness. *I AM* experiencing deeper insights. *I AM* connected with higher levels of energy and a vibration of love that has a practical Earthly application. I know how to apply this love in my life, in my relationships and in my group interaction in the world. *I AM* able to put LOVE in ACTION. *I AM* ready to Be of Service. I know that Everything is Perfect. . . Everything is in Divine Order. SO BE IT! Only that which is for my higher good can come from using these frequencies.

Manifestation

I AM grateful for the easy and graceful transformation that is occurring in all of my bodies as a result of using these frequencies.

I AM bringing all the levels of my brain and mind into harmony. *I AM* able to maintain a level of one point focus of concentration. *I AM* able to bring all the levels of the various aspects of the brain and mind into harmony for a brief moment so I can manifest. *I AM* free from the self-imposed blocks which prevent manifestation. *I AM* free from the misuse of manifestation. *I AM* willing to reawaken, to open to my inner sensitivity and consciousness. *I AM* willing to use my higher Self, my highest ability to help others joyfully with all my heart. *I AM* free from all viruses held in my spine. *I AM* enjoying perfect health in the area of my heart. *I AM* experiencing a sense of purpose that manifests in my physical body. *I AM* receiving these frequencies which are emitted as part of the solar spectra and the energy from the Center of galaxies. *I AM* in the understanding that what manifests in my world comes from my Being first, then Doing, then Having. *I AM* understanding more of my purpose with an awareness of my BEINGNESS. *I AM* experiencing a sense of calm and peace with energy behind it resulting in an awareness of the role my higher guides play. *I AM* now able to work harmoniously in a group for a higher purpose. I know that Everything is Perfect. . . Everything is in Divine Order. SO BE IT!

Only that which is for my higher good can come from using these frequencies.

INTENTIONS FOR
THE MENTAL / EMOTIONAL SERIES
Spacelight

I AM grateful for the easy and graceful transformation that is occurring in all of my bodies as a result of using these frequencies.

I AM a Being of Light. *I AM* surrounded by Light. *I AM* declaring this space to be open *only* to the Light. *I AM* demanding that every being that is lost in this place turn toward the Light and return home to waiting friends and family. I know that Everything is Perfect. . . Everything is in Divine Order. SO BE IT! Only that which is for my higher good can come from using these frequencies.

Safe Environment

I AM grateful for the easy and graceful transformation that is occurring in all of my bodies as a result of using these frequencies.

I AM enriched by breathing in noble gases, (especially Argon and Krypton), that were present in sufficient quantities in the atmosphere at a time when there was no pollution. As my third and sixth chakras, (the seat of my power and spiritual centers), align into a state of resonance. *I AM* able to experience a deeper sense of my connection to the Soul of Humanity and to the connection between the earth, the sun and other energies. *I AM* living in an environment where all kingdoms on Earth may flourish and all living things are supported, so they can accomplish their Soul Mission. I know that Everything is Perfect. . . Everything is in Divine Order. SO BE IT! Only that which is for my higher good can come from using these frequencies.

Clarity & Focus

I AM grateful for the easy and graceful transformation that is occurring in all of my bodies as a result of using these frequencies.

I AM increasing my awareness of my inner guidance. *I AM* receiving visual information. *I AM* now able to stay mentally focused. I know my purpose here on the planet at this time. *I AM* remaining in the moment, which is my key to solving many of my issues and my key to being happy right now. *I AM* able to see the connection between my past lives and my present life with the help of these frequencies that provide access to information related to my currant life situation. *I AM* willing to bring these experiences into my own knowing. *I AM* able to remain focused on what I want on this physical plane. *I AM* willing to receive information from my multidimensional Self. *I AM* coming into the knowledge of my Soul purpose in this incarnation. *I AM* clearly focused on my purpose. *I AM* focused in the Eternal Now. I know that Everything is Perfect. . . Everything is in Divine Order. SO BE IT! Only that which is for my higher good can come from using these frequencies.

Communication

I AM grateful for the easy and graceful transformation that is occurring in all of my bodies as a result of using these frequencies.

I AM a loving teacher. *I AM* a loving student. As a teacher, *I AM* able to see through a student's perspective, to see the learning blocks and shift them, understanding them in a new way. *I AM* improving my communication skills. *I AM* willing to let go of the manner in which I see myself. *I AM* open and honest about my experiences both internal and external. *I AM* willing to change and receive new energies. *I AM* connected and in attunement as a teacher and as a student. *I AM* experiencing cellular regeneration in a "youthening" manner because of greater absorption of the minuscule amounts of Xenon stimulation in the Earth's atmosphere. I know that Everything is Perfect. . . Everything is in Divine Order. SO BE IT! Only that which is for my higher good can come from using these frequencies.

Choice

I AM grateful for the easy and graceful transformation that is occurring in all of my bodies as a result of using these frequencies.

I AM experiencing more energy in my master hormones, particularly DHEA, which improves my brain/body relationship. *I AM* looking at all the deeper issues involved. *I AM* serving others, working with others and loving them. *I AM* reducing my exposure to television and all other addictive patterns. *I AM* raising my level of consciousness. *I AM* able, on the level of the Soul Self, to choose, to find the path and to know it. I know that Everything is Perfect. . . Everything is in Divine Order. SO BE IT! Only that which is for my higher good can come from using these frequencies.

Memory

I AM grateful for the easy and graceful transformation which is occurring in all my bodies as a result of using these frequencies.

 I AM attuned to several levels of past life connection particularly with the Animal Kingdom. *I AM* aware of the strengthening of the link between the astral and emotional bodies. *I AM* experiencing a heightened dream state and a deeper awareness of past life connection to the animal kingdom in any form. *I AM* free from all skin conditions. *I AM* able to transform energy from one level to another. *I AM* willing to work with such levels of transformation, to switch emotional energy into mental energy for improved memory and recognition. *I AM* communicating with the animal kingdom easily. *I AM* listening easily to the animal's message, to my animal guide, to my animal totem. I know that Everything is Perfect. . . Everything is in Divine Order. SO BE IT! Only that which is for my higher good can come from using these frequencies.

Public Speaking

I AM grateful for the easy and graceful transformation that is occurring in all of my bodies as a result of using these frequencies.

 I AM willing to receive a new sense of my speaking capability. *I AM* able to see myself through the eyes of others with love. *I AM* able to go through the process of slowing my thought rates, which enables my voice to transform energy from thought. *I AM* open to channeling from the highest source at my choice. *I AM* joyfully connected to my audiences in love and harmony. *I AM* able to hear my guides. *I AM* meditating more easily. *I AM* recognizing my influence in the Universe and how it responds to me. I know that Everything is Perfect. . . Everything is in Divine Order. SO BE IT! Only that which is for my higher good can come from using these frequencies.

Body Wisdom

I AM grateful for the easy and graceful transformation that is occurring in all of my bodies as a result of using these frequencies.

I AM releasing old belief systems about my body that no longer serve me. *I AM* focusing on all that is good and right about my body. *I AM* loving my body *just the way it is now. I AM* grateful for all the parts of my body that serve me perfectly today. *I AM* receiving with an open, loving heart, and have the wisdom to make ideal food choices. I enjoy food that turns into beauty and good health. *I AM* attaining and maintaining my ideal weight. I feel confident, refreshed and full of positive energy. *I AM* becoming even more perfect as I provide my body with love and support. I know that Everything is Perfect. . . Everything is in Divine Order. SO BE IT! Only that which is for my higher good can come from using these frequencies.

Addictions

I AM grateful for the easy and graceful transformation that is occurring in all of my bodies as a result of using these frequencies.

I AM releasing recognized addictions and those of which *I AM* not consciously aware. *I AM* free from cravings for substances at the cellular level. *I AM* filling this space with life sustaining harmony. *I AM* transforming my old unconscious reaction to life, to a conscious response based on wisdom, oneness and love. *I AM* living a happy life. I know that Everything is Perfect. . . Everything is in Divine Order. SO BE IT! Only that which is for my higher good can come from using these frequencies.

Emotional Release

I AM grateful for the easy and graceful transformation that is occurring in all of my bodies as a result of using these frequencies.

I AM experiencing an inner state of peace that allows me to find the cause of my pain. *I AM* free from my physical and emotional pain. *I AM* experiencing a vertical flow of energy which is contributing to the overall balance of my chakra system. *I AM* enjoying a state of internal harmony that tends to help me find balance and harmony in groups. *I AM* balanced in my body. *I AM* expressing my emotions so I can be free. *I AM* imbued with an inner sense of peace. I know that Everything is Perfect. . . Everything is in Divine Order. SO BE IT! Only that which is for my higher good can come from using these frequencies.

Brain / Body

I AM grateful for the easy and graceful transformation that is occurring in all of my bodies as a result of using these frequencies.

I AM experiencing a deeper attunement between my physical and mental bodies. *I AM* open. *I AM* free of judgment and blocks in my thinking. *I AM* experiencing a shift of energy that is enabling me to change my point of view. *I AM* in a perfect acid / alkaline balance. *I AM* conscious and aware and open to new dietary principles. *I AM* eating more appropriately and making perfect food choices. I know that Everything is Perfect. . . Everything is in Divine Order. SO BE IT! Only that which is for my higher good can come from using these frequencies.

#21

I AM grateful for the easy and graceful transformation that is occurring in all of my bodies as a result of using these frequencies.

I AM bringing new Light and well being with the vehicle of Light to each cell. *I AM* focused on understanding the Truth. *I AM* aware of any issues which may be preventing the raising of my consciousness. *I AM* acknowledging and releasing any causes of imbalance that are triggered and brought to the surface for me. *I AM* ready to share my process of clearing. *I AM* a Being of Light and Joy. I know that Everything is Perfect. . . Everything is in Divine Order. SO BE IT! Only that which is for my higher good can come from using these frequencies.

INTENTIONS FOR THE PHYSICAL SERIES

Prana

I AM grateful for the easy and graceful transformation that is occurring in all of my bodies as a result of using these frequencies.

I AM benefiting on all levels from a perfect balance of oxygen which is an essential element of the primary double helix structure of the DNA and RNA molecules. *I AM* accelerating the release of clogged emotions. *I AM* breathing LIFE FORCE into my body. *I AM* breathing out, expelling what no longer serves me. *I AM* creating a vacuum to be filled with LIFE, LIGHT and LOVE. *I AM* rising above the appearance of the opposite, and realizing ONENESS in the Breath of Life. I know that Everything is Perfect. . . Everything is in Divine Order. SO BE IT! Only that which is for my higher good can come from using these frequencies.

Hair

I AM grateful for the easy and graceful transformation that is occurring in all of my bodies as a result of these frequencies.

I AM able to transfer energy through my spine and my energy centers in a vertical direction. *I AM* experiencing perfect balance of my hormones. *I AM* enjoying perfect circulation in my skin, particularly in my scalp. *I AM* growing my hair on my head. *I AM* recirculating my sexual energy. *I AM* moving it where it is appropriately felt and balanced. *I AM* perfectly balanced internally. *I AM* getting along with others and understanding others. *I AM* able to work in groups appropriately and harmoniously. I know that Everything is Perfect. . . Everything is in Divine Order. SO BE IT! Only that which is for my higher good can come from using these frequencies.

Circulation

I AM grateful for the easy and graceful transformation which is occurring in all my bodies as a result of using these frequencies.

I AM cleansing and energizing my arteries, veins and lymphatic system. *I AM* opening my arteries and veins that carry nutrients into my body and toxins out of my body. With the help of these frequencies, *I AM* stimulating my lymphatic system and helping it fulfill its important function of keeping my body well and guarding against the invasion of foreign bodies. *I AM* delighting in an increase of energy. *I AM* enjoying benefits to my skin, hair and to all aspects of my outer body because of the greatly improved circulation and more efficient delivery of nutrients to those parts of my body. *I AM* experiencing a deep sense of physical well being, that is reflected in my emotional and mental bodies as well. I know that Everything is Perfect. . . Everything is in Divine Order. SO BE IT! Only that which is for my higher good can come from using these frequencies.

Cell Rejuvenation

I AM grateful for the easy and graceful transformation that is occurring in all of my bodies as a result of using these frequencies.

I AM remembering, at my cellular level, my cells' original function, as well as how to gain nourishment for growth. *I AM* creating a perfect climate for cellular rejuvenation and growth. *I AM* quickly releasing foreign bodies from my system. *I AM* feeling enhanced on all levels. *I AM* rejuvenating peacefully and in Divine Time. I know that Everything is Perfect. . . Everything is in Divine Order. SO BE IT! Only that which is for my higher good can come from using these frequencies.

Bones

I AM grateful for the easy and graceful transformation that is occurring in all of my bodies as a result of using these frequencies.

I AM able to remain in the NOW. *I AM* producing bone marrow of the highest quality blood cells. Focused in the NOW, *I AM* looking for ways in which to express joy, happiness, and abundance. *I AM* in a climate, in which my body reproduces itself in such a way that each cell and membrane performs its task. *I AM* basking in that feeling of well being, knowing that all is well, functioning smoothly and harmoniously. *I AM* projecting feelings of joy and abundance into my future and my past, thereby pulling into my present all experiences which resonate with that feeling of abundance and well being. I know that Everything is Perfect. . . Everything is in Divine Order. SO BE IT! Only that which is for my higher good can come from using these frequencies.

Hearing

I AM grateful for the easy and graceful transformation that is occurring in all of my bodies as a result of using these frequencies.

I AM balanced and healed at all levels. *I AM* free from all hearing disorders and infections. *I AM* joyful and balanced as all hearing disturbances are reduced and my inner ear is normalized. *I AM* able to hear my inner guidance. *I AM* experiencing a feeling of well being as my inner and outer ear are balanced. I know that Everything is Perfect. . . Everything is in Divine Order. SO BE IT! Only that which is for my higher good can come from using these frequencies.

Hydration

I AM grateful for the easy and graceful transformation that is occurring in all of my bodies as a result of using these frequencies.

I AM, through absorption of these frequencies, balancing the energy moving up and down in my body, allowing the Kundalini energy arising in my spine to move properly. *I AM* experiencing an increase of water absorption in my body which improves the quality of my skin and enhances the absorption of nutrients from other liquids. I know that my cells are becoming more discriminating, taking only what is nutritious and rejecting what is not. *I AM* deepening my sense of internal revelation. *I AM* able to understand concepts that were previously unclear to me. *I AM* increasing the connection of my cellular structures, my Soul and my brain. *I AM* experiencing an increase in the conductivity of my connective tissue, that provides stimulation and benefits to my kidneys. I know that Everything is Perfect. . . Everything is in Divine Order. SO BE IT! Only that which is for my higher good can come from using these frequencies.

Compassionate Heart

I AM grateful for the easy and graceful transformation that is occurring in all of my bodies as a result of using these frequencies.

I AM able to know love, to accept love and to give it freely without expecting or needing anything in return. *I AM* absorbing Vitamin C. *I AM* producing Vitamin C that is essential for the flexibility of my arteries. *I AM* feeling a perfect balance of the magnetic and electrical nerve impulses. *I AM* able to move oxygen in and out of my cells. *I AM* able to take in and release on a constant basis all the ingredients necessary for my life. *I AM* able to love, receive love and to let go. *I AM* able to transfer energy through my spine to all my energy centers. *I AM* producing hormones in my glands with perfect balance so that my body functions optimally. *I AM* balanced within. *I AM* enjoying easy companionship, cooperation and mutual understanding. *I AM* experiencing an inner revelation regarding earlier choices I have made, that led to actions which may not have been beneficial to me. I know that Everything is Perfect. . . Everything is in Divine Order. SO BE IT! Only that which is for my higher good can come from using these frequencies.

Pancreas

I AM grateful for the easy and graceful transformation that is occurring in all of my bodies as a result of using these frequencies.

I AM entering into a period of self-acceptance and self-love. *I AM* grateful for my pancreas that is working perfectly and is secreting the digestive juices for perfect assimilation. *I AM* delighting in my body that is full of joy and in my organs that sing in harmony. *I AM* a perfect body weight NOW. I know that Everything is Perfect. . . Everything is in Divine Order. SO BE IT! Only that which is for my higher good can come from using these frequencies.

Perfect Veins

I AM grateful for the easy and graceful transformation that is occurring in all of my bodies as a result of using these frequencies.

I AM able, in my nervous system, to increase the blood flow throughout my body as well as around my heart. *I AM* absorbing the amino acids taurine and lysine through my veins, and thus, through these frequencies, *I AM* able to repair my veins when under stress. *I AM* experiencing a perfect flow of oxygen in and out of my cells. *I AM* rekindling ideas in my consciousness that need to be re-circulated. *I AM* recreating aspects in my world that have been valuable to me. *I AM* able to project these precious gifts to the world. *I AM* able to take in and release all that is necessary for my life. *I AM* able to love appropriately, to love, to receive love and to let go. I know that Everything is Perfect. . . Everything is in Divine Order. SO BE IT! Only that which is for my higher good can come from using these frequencies.

Stability & Stillness

I AM grateful for the easy and graceful transformation that is occurring in all of my bodies as a result of using these frequencies.

I AM willing to reach for, and accept balance, harmony and coherence with my heart and my entire body. *I AM* seeking and achieving a state of inner peace. *I AM* enabling energy to flow freely and perfectly through my spine and through my energy centers. *I AM* realizing that portions of my earlier life and the manner in which my energy was spent, may not have been helpful to myself or to the greater good. *I AM* now in a state of perfect balance and harmony with my self and others, to the extent that group activities are fun and fulfilling. I know that Everything is Perfect. . . Everything is in Divine Order. SO BE IT! Only that which is for my higher good can come from using these frequencies.

Vitality & New Health

I AM grateful for the easy and graceful transformation that is occurring in all of my bodies as a result of using these frequencies.

I AM absorbing new information easily. *I AM* visualizing things with a new perspective. *I AM* taking in what I need. *I AM* releasing all components that are no longer necessary in my life, whether it is relationships, organizational patterns, or habits. *I AM* able to love appropriately, that is, to love, to receive love and to let go. *I AM* enhanced in the area of my small intestine. *I AM* strengthened in the area of my immune system particularly in my T cells. *I AM* relearning the ability to move oxygen in and out of my cells. With the help of these frequencies, *I AM* creating from raw materials the necessary digestive components, particularly hydrochloric acid. *I AM* enjoying perfect health and perfect vitality. I know that Everything is Perfect. . . Everything is in Divine Order. SO BE IT! Only that which is for my higher good can come from using these frequencies.

Highlights of Tapes

HIGHLIGHTS OF THE
SOUND WAVE ENERGY TAPES

These tapes are not FDA approved. They are used for personal research and educational purposes only. They are recorded in real time and cannot be duplicated. Duplication may alter the effectiveness and results.

FOUNDATION SERIES

CHAKRA #1 BASE OR ROOT
Elimination functions - bowel, bladder. Grounding to earth.

CHAKRA #2 SACRAL, SPLEEN
Reproductive organs, life energy; male and female. Loving relationship with Self and others. Balance.

CHAKRA # 3 SOLAR PLEXUS
Digestion - liver, stomach, pancreas, small intestine, adrenal glands. Helps to accept and digest challenges of life. Seat of emotions and power. Improves assimilation.

CHAKRA # 4 HEART, THYMUS
Blood, circulation, lymphatics. Expresses love and wisdom. Strengthens immune system. Includes growth hormone.

CHAKRA # 5 THROAT
Thyroid, parathyroid, vocal area, respiration, lungs. Regulates metabolism. Helps to speak our truth. Guides peace and trust to creative expression.

CHAKRA #6 PITUITARY
Opens third eye, clairvoyance. Eyes, nose, ears, teeth, lower brain. Regulates fluids acid/alkaline. Produces growth hormones.

CHAKRA #7 CROWN
(Also Chakras 8 & 9)
Pineal gland connection. Brain and nervous system. Connection to GOD.

RELAXATION & CALMNESS
Letting go. Works with DNA. Balances mind and heart. Removes resistance and stress.

LOVE
Increases feeling of well-being; unconditional love for Self and others. Golden Light.

HIGHER CONSCIOUSNESS
Releases blocks, fear, doubt, disharmony. Raises vibrations to Christ Consciousness.

BRAIN, COURAGE & PROSPERITY
Grounding; improves decision-making. Synchronizes both hemispheres of brain. Courage to accept abundance.

SPIRITUAL SERIES

THYMUS
Master gland for immune system. Seat of unconditional love. Reverses aging. Stay young.

TRANSITION
Removes limitations. Facilitates the acceptance of new things and miracles.

HIGHLIGHTS

(continued)

MERKABA
Peace and serenity; allows deep
meditative state. Creates 55 feet of
impenetrable vortex of protection
around us.

ENERGY, DNA & ENZYMES
Helps manifest enzymes. Balances chakras 2,4,6 and 8.
Strengthens the digestive system
and energizes the body. Improves
the ability to love and bring love
into any situation. Assists in getting rid of parasites and micro-
organisms. Good for athletes.
Allows future parents to choose
from their DNA the characteristics
to transmit to their child.

PATIENCE
Connects with the plant kingdom.
Develops patience. Breaks the
attachment to attention and the
need for approval. Helps in
grounding. Helps relieve symptoms of Chronic Fatigue and
Fibro-Myalgia. Maintains strong
lungs.

ONENESS
Enhances feelings of Universal
Love and Oneness with mankind
and the Universe.

PURIFICATION
Helps to release unconsciously
gathered entities. Cleanses and
energizes the blood. Assists in

PURIFICATION (continued)
absorbing energy from the sun,
air, water and food. Helps us
understand the proper use of
energy. Assists with recovery after
surgery.

GATEWAY
Receive new energies and integrate them. Helps us revise our
values and release material attachment. Increases sense of
inner freedom.

IMMORTALITY
Helps stimulate higher endocrine
function in the pineal and pituitary glands. Helps to understand
the Will of the Soul and the Will
of God. Influences cellular regeneration.

CONNECTIONS
Assists in balancng the ego. Increases the healer's ability to heal
themselves and others. Supports a
deeper communication with
Angels. Promotes self forgiveness,
clarity, awareness, and understanding. Facilitates a strong
attunement with extra-terrestrial
sources. Helps clear states of
fatigue and difficulties tuning to a
higher vibration. Appears to
supplement and strengthen our
psychic abilities and higher God
connections.

SPIRITUAL SERIES (continued)

SHIFTING CONSCIOUSNESS
Helps attune to humanity's Greater Purpose and supports quantum leap shifts. Encourages harmonious interactions with others. Raises consciousness. Assists in putting love in action and promotes joyful service. Helps strengthen the immune system.

MANIFESTATION
Increases the ability to manifest. Helps instill a sense of purpose and a better understanding of our "beingness." appears to enhance liver function and ease heart conditions. Promotes awareness of our Higher Guides.

MENTAL & EMOTIONAL SERIES

SPACELIGHT
Brings light into a space. Helps clear negativity.

SAFE ENVIRONMENT
Allows the third Chakra (Energy Center) and the Sixth Chakra to become aligned, and re-establishes a safe environment.

CLARITY & FOCUS
Helps us receive visual information, discover our purpose, achieve clarity, maintain focus and "BE" in the moment.

COMMUNICATION
Improves communication, promotes an innter student/teacher relationship. Seems to improve cellular regeneration.

CHOICE
Assists those stuck in service to self; brings in a larger level of the Soul Self.

MEMORY
Assists with the attunement of past life connections, especially with the animal kingdom; heightens dream state, improves breathing through oxygenation, improves memory, recognition, and learning. Lessens the effects of Herpes and Shingles.

PUBLIC SPEAKING
Helps ease the fear of public speaking. Influences our ability to hear our guides. Assists in meditation.

BODY WISDOM
Addresses fears and judgment regarding body size; promotes total self-acceptance.

ADDICTIONS
Releases substance addictions. Removes cravings at the cellular level. Sustains life and harmony in the cells. Releases and transforms addictive patterns.

EMOTIONAL RELEASE
Helps modulate pain reaction and find the underlying cause of the pain. Supports the natural process of opening up and releasing through crying. Stimulates certain substances in the brain that restore a state of inner peace.

BRAIN / BODY
Appears to bring a shift in brain chemistry. Balances acid/alkaline levels.

#21 - LIGHT TRANSFORMATION

Holds "New Light" and generates a feeling of wellbeing. Helps to focus on the understanding of Truth; assists in bringing up old issues for review and release.

PHYSICAL SERIES

PRANA
Appears to allow more oxygen into the body. Assists the lungs, breathing, and athletic performance. Encourages emotional clearing.

HAIR
Helps re-circulate sexual energy. Assists circulation in the skin, including the scalp, which helps to re-grow the hair.

CIRCULATION
Cleans and energizes the circulatory and lympthatic system.

CELL REJUVENATION
Promotes a harmonious relationship among cells. Creates the proper environment for cells to grow and reproduce. Assists in creating an environment in which viruses, "bad" bacteria, and fungus cannot survive.

BONES
Appears to help restores healthy bones and joints. Promotes growth of healthy cells in the bone marrow. Keeps our focus on the present.

HEARING
Normalizes the inner ear with balance and joy. Assists with hearing disturbances and loss of hearing. Clears infections of the inner ear. Assists in hearing one's guidance.

HYDRATION
Assists the body in water absorption and the absorption of nutrients from other liquids. Increases conductivity of the connective tissue. Is helpful to the kidneys and skin, appears to help those with multiple sclerosis.

PHYSICAL SERIES (cont'd)

COMPASSIONATE HEART
Helps us know and accept love and give it freely. Influnces vitamin C absorption and production. Strengthens the entire body's ability to move oxygen in and out of cells.

PANCREAS
Normalizes pancreatic function.

PERFECT VEINS
Influences the nervous system for a better veinous blood flow return. Helps to repair veins when under stress. Helps the body move oxygen in and out of the cells. Rekindles desire to recirculate into the world precious gifts that have been valuable to us.

STABILITY & STILLNESS
Helps enhance the coherence of the heart. Helps restores an inner state of peace by helping to increase certain substances in the brain.

VITALITY & NEW HEALTH
Helps to strengthen the small intestine. Bolsters the immune system. Appears to create the necessary components for good digestion. Assists in absorbing new information and seeing things in a new way.

Product Information

PRODUCT INFORMATION
TAPES & CDs
See Chapter Ten for recommended combinations.

Foundation Series, 11 tapes or 12 CDs   **$369.00**
Chakra (Energy Center) Set, plus   + 11.00 S/H
Relaxation & Calmness; Love;
Higher Consciousness;
Brain, Courage & Prosperity
Book, *Return to Harmony* by Nicole LaVoie
Video by Nicole LaVoie

Spiritual Series, 12 tapes or CDs   **$369.00**
Thymus; Transition; MerKaBa;   +11.00 S/H
Energy, DNA & Enzymes;
Patience; Purification; Oneness;
Gateway; Connections; Shifting Consciousness,
Manifestation; Immortality

Mental/Emotional Series, 12 tapes or CDs   **$369.00**
Spacelight; Safe Environment; Clarity & Focus;   +11.00 S/H
Communication; Choice; Memory;
Public Speaking; Body Wisdom; Addiction;
Emotional Release; Brain/Body; #21

Physical Series, 12 tapes or CDs   **$369.00**
Prana; Hair; Circulation; Cell Rejuvenation;   +11.00 S/H
Bones; Hearing; Hydration; Compassionate Heart;
Pancreas; Perfect Veins; Stability & Stillness;
Vitality & New Health

Total for the Four Series   **$1,476.00**
    +30.00 S/H

Book, *Return to Harmony* by Nicole LaVoie   **$13.99**
(Additional books $1.50 S/H)   +3.00 S/H

Video, "Balance the Bodies with Sound Waves"   **$10.00**
By Nicole LaVoie; 75 minutes   +3.00 S/H

Voice Test for SoulNote (must be done IN PERSON)   **$125.00**

NOTE: The audiotapes are produced on top quality cassettes and are recorded in real time, using specially developed equipment to ensure accuracy to four decimal places. *Copying tapes or CDs will alter the frequencies and the results.*

INDIVIDUAL TAPES

The six recordings listed below may be purchased individually. Each recording is **$36.00** plus S/H (see below).

Relaxation & Calmness
Prana
SpaceLight
Hair
Safe Environment

Lean Body
Please Note: The Lean Body recording MUST be used in conjunction with the Foundation Series. If you do not have the Foundation Series, you must order this before ordering the Lean Body recording.

Shipping & Handling:
Series:
1 SERIES=$11.00 • 2 SERIES=$18.00 • 3 SERIES=$25.00 • 4 SERIES=$30.00

Individual Tapes:
One to four individual tapes, $4.00 S/H. Five tapes, $5.00; six tapes, $6.00; seven to ten tapes, $7.00; and eleven tapes, $11.00.

NOTE: To obtain proper results with the SWE frequencies, you may acquire any one or all of the four Series, but THE FOUNDATION SERIES IS A PREREQUISITE TO ALL OTHER SERIES. You must work with the Foundation Series to provide a good foundation to your four bodies (Spiritual, Mental, Emotional, and Physical).

To order, contact:
Sound Wave Energy
PO Box 3969
Pagosa Springs, CO 81147 USA

Phone 970-264-6469
Fax 970-264-2696
1-888-267-2309
email: swe@cslswe.com
web site: www.swe@cslswe.com

Visa, Mastercard, Discovery, American Express Accepted

Make checks payable to Sound Wave Energy

Orders are shipped via US Postal Service the same or next day.

Prices subject to change without notice.